SYNCHRONOUS PLANNED MAINTENANCE

The Business of Constraint Management

ROBERT S. HILLIGOSS

Copyright © 2022 Robert S. Hilligoss.

All rights reserved. No part of this book may be used or reproduced by any means, graphic, electronic, or mechanical, including photocopying, recording, taping or by any information storage retrieval system without the written permission of the author except in the case of brief quotations embodied in critical articles and reviews.

Archway Publishing books may be ordered through booksellers or by contacting:

Archway Publishing
1663 Liberty Drive
Bloomington, IN 47403
www.archwaypublishing.com
844-669-3957

Because of the dynamic nature of the Internet, any web addresses or links contained in this book may have changed since publication and may no longer be valid. The views expressed in this work are solely those of the author and do not necessarily reflect the views of the publisher, and the publisher hereby disclaims any responsibility for them.

Any people depicted in stock imagery provided by Getty Images are models,
and such images are being used for illustrative purposes only.
Certain stock imagery © Getty Images.

Scripture taken from the King James Version of the Bible.

ISBN: 978-1-6657-0734-3 (sc)
ISBN: 978-1-6657-0733-6 (e)

Library of Congress Control Number: 2021910602

Print information available on the last page.

Archway Publishing rev. date: 11/16/2022

Contents

Preface vii
Introduction xi

Chapter 1: Issues and Opportunities 1
Chapter 2: The Paradigm Effect 4
Chapter 3: Maintenance 101 6
Chapter 4: Dollars and Sense 13
Chapter 5: Managing by the Numbers 17
Chapter 6: Constraints, Barriers, Roadblocks and Delays 21
Chapter 7: Preventive Maintenance 27
Chapter 8: Maintenance Spare-Parts Control 36
Chapter 9: Priorities 50
Chapter 10: Organizational Structure 55
Chapter 11: Management Reports 68
Chapter 12: Core Business Focus 71
Chapter 13: Worker Productivity 75
Chapter 14: Operator Machine Care 81
Chapter 15: High-Tech Manufacturing and Plant Maintenance 85
Chapter 16: Machine History 90
Chapter 17: Education and Training 99
Chapter 18: The Evolution of Maintenance Dispatching 101
Chapter 19: Maintenance Hourly Employee Survey 109
Chapter 20: System Concepts 101 124
Chapter 21: The Challenges of Positive Change 131
Chapter 22: Common-Sense Indices of Business Performance 134

About the Author 149

Preface

Synchronous Planned Maintenance offers a comprehensive focus on managing the business aspects of plant maintenance; its effect on the goals and objectives of the business enterprise; and how plant maintenance can be leveraged to the economic advantage of the larger business enterprise. I have had the opportunity and privilege of associating with many excellent maintenance professionals, and I am amazed by the skills and talents some of these people have demonstrated in managing the maintenance function—more often than not under adverse circumstances over which they have little or no control. Unfortunately, many of these people are not managing effectively due to the constraining influences of the plant maintenance environment.

This book is dedicated to these supervisors and managers, who, if it were made possible, could transform plant maintenance, becoming enablers to the success of the larger business enterprise. The purpose of this book is to influence those key executives and senior managers who have it in their power to make this outcome possible. Now, a business book about plant maintenance may not rank very high on a busy executive's list of things to read. But they are the ones who can make the decisions and take the actions to transform plant maintenance into American industry's next competitive edge.

In writing this book, I have a concern that requires some explanation. Knowledge and understanding in any field of endeavor evolves over a period of time, not only from our own personal experience but also as a result of association with other people and exposure to their thoughts and ideas. During my forty-plus years of involvement in plant engineering and maintenance at General Motors and EDS, I have had the opportunity to meet and work with many knowledgeable maintenance professionals. It was my privilege to be involved with GM education and training—and later with GM's advance engineering staff, plant engineering and maintenance group—in the development and presentation of training programs that were attended by a large number of maintenance managers and supervisors.

Over the years, I have been a frequent speaker and participant in national and international conferences, workshops, and seminars where maintenance issues are discussed and thoughts and ideas are freely exchanged. In addition, I have a library of maintenance books and publications by a host of authors. I mention these things to show how other people may have in some way directly or indirectly contributed to this book. I want to avoid the possibility of unintentionally failing to give proper credit to someone. If this should occur, it will be corrected in future publications.

I do want to recognize four people who have influenced my thinking, although they never knew it. First is Mr. Russ Ackoff, who was dean of the Wharton School of Business. He pulled back the curtain just a wee bit to provide an intellectual view of the world of systems and systems thinking. Most people are simply not aware of the dynamic role of systems as the primary engine that drives almost all organizational activities, including plant maintenance.

I also thank Eli Goldratt for writing his book *The Goal*. Mr. Goldratt used so many everyday common-sense examples to make a complex subject (variation and dependencies) so simple even I could understand it. It provided insight into the role of systems and systems thinking in the success (or failure) of manufacturing organizations.

Looking at the attributes and characteristics of quality, we should recognize it as the ultimate goal of the organization. Mr. Phillip Crosby provided many practical examples for applying system concepts and systems thinking to bring about quality in the manufacturing environment.

Finally, I want to thank Mr. Joel Barker for introducing me to the concept of paradigms in his book *Discovering the Future*. Understanding the concept of paradigms made writing this book possible. Mr. Barker explains how paradigms can make highly intelligent and capable people become completely blinded to what otherwise would be so obvious. He uses several compelling examples of major international companies that had dominated their marketplace in the past losing that marketplace to their competition. Their past success was their downfall, because their paradigms blinded them to what was coming in the future.

We live in a world of paradigms, and understanding this makes it worthwhile to try to communicate with busy executives and senior managers who may have become just a little bit blinded to some very obvious things that can be vital to their future success or failure.

Although Russ, Eli, Phil, and Joel have had the most to do with shaping my own thinking, there were many others. I realize it's not practical, or even possible, to recognize everyone personally, but I can use this opportunity to express my sincere thanks and appreciation for each and every individual in that legion of anonymous contributors to this book.

Introduction

Synchronous planned maintenance (SPM) is a concept that's focused on organizational structure, systems, processes, policies, procedures, and practices. These are the tools used by organizations to manage their operations. Dysfunction in these management controls is one of the primary causes of work constraints in the plant maintenance workplace. These continuing work constraints result in excessive maintenance costs and associated production costs and losses, as seen over the past fifty to sixty years or more.

In too many cases, these management tools have become dysfunctional due to flawed initial design or changing operational environment. This is especially true in discrete manufacturing plants. When properly designed and implemented, SPM can reduce maintenance and production costs as well as have a positive impact on the bottom-line profits of a manufacturing plant.

The objective of this book is to bring about a paradigm shift in the way people think about their plant maintenance operations and focus their attention on the multitude of opportunities for major cost savings and cost avoidance. The greater part of these benefits will occur outside of the maintenance department, in other areas of plant organization. Valuable information is provided for senior managers and other key decision-makers about how to transform their plant maintenance operations from a burden cost to a major contributor to the financial objectives of the larger business enterprise.

Many of the traditional concepts of plant maintenance have become obsolete in the era of modern manufacturing. There is a need for a paradigm shift where the concept of plant maintenance assumes new dimensions and the plant maintenance function is transformed from a departmental activity into a more global plant-wide set of interrelated and interdependent processes.

In the average discrete manufacturing plant, the total cost of maintenance (TCM) is much greater than the maintenance budget costs (MBC). This is due to additional costs that are referred to as maintenance-related costs (MRC). MRC are the upstream and downstream costs and losses that are incurred outside of the maintenance department in other areas of the plant operation, as direct or indirect consequences of maintenance-related problems.

Historically, the TCM in most plants has been excessively high—but that can be corrected, and the sooner the better. The ultimate objective of SPM is to help facilitate the changes needed to achieve the minimum TCM.

SPM creates an enterprise-wide approach to plant maintenance, helping to resolve maintenance-related problems that have plagued manufacturing plants for far too long, using the following key elements:

- **Synchronism**—A set of interrelated and interdependent parts all work together simultaneously.
- **Planning**—Design characteristics define the role of the parts and govern how the parts all work together.
- **Maintenance**—Preservation of physical plant assets are a top priority.

This book presents a comprehensive focus on many of the things that can and usually do go wrong to adversely affect plant maintenance operations. There's an old saying that "You can't fix it unless you know what's broken." A focused close-up view clearly exposes what's broken.

In virtually every manufacturing plant, there will be some level of dysfunction in organizational structure, systems, processes, policies, procedures, and practices that result in on-the-job roadblocks, barriers, and delays. These constraints reduce the effectiveness, efficiency, and utilization of the skilled maintenance workforce. This results in extended machine downtime, direct labor idleness and inefficiency, lost production, overtime, premium freight, late shipments, and other direct and indirect production costs and losses.

Obviously, nobody likes to think that these kinds of problems are of significance in their own plant. In fact, some may even become defensive at such a suggestion. However, experience has shown that these (undocumented) work constraints have become major problems in many if not most discrete manufacturing plants. On-the-job work constraints incurred by a plant's skilled maintenance workforce can add significantly to the costs of production, stifling the best efforts of even the most dedicated organizations and keeping them from achieving their potential.

Organizational structure, systems, processes, policies, procedures, and practices are management tools; if problems occur, they are management problems requiring management solutions. In many cases, these problems reflect complex systemic issues, and solutions may require changes in traditional organizational structure and areas of responsibility. In these cases, the active involvement of senior management is essential, because they are the only ones who are in a position to make changes both within and across organizational boundaries.

Positive change in these areas of dysfunction will bring about more synergy in the customer-supplier relationships that exist between individuals, work groups, departments, plants, etc. The material presented here will arm managers with a better understanding about these areas of dysfunction, how they reduce operational effectiveness, and what it takes to bring about positive change. Given the opportunity, SPM can facilitate an era of meaningful, measurable improvement where it counts the most: at the financial bottom line.

I will share my professional insight and lessons learned during a career in plant maintenance that has spanned nearly forty-five years, working in small, medium, and large manufacturing plants—in the beginning as a maintenance skilled tradesman, and many years and experiences later as a senior manufacturing consultant and recognized authority in the concepts of planned maintenance.

Chapter 1
Issues and Opportunities

After World War II, with the assistance of the United States, Japan began the effort of rebuilding shattered factories and restoring its manufacturing capabilities. The experience of having to start over and create an industrial base from the ground up provided a unique opportunity to develop profoundly different approaches to manufacturing operations. The Japanese owe much of the credit for their success to an American, Mr. Edward Deming, whose work with the Japanese in the 1950s and beyond instilled the quality ethic into the culture of Japanese manufacturing.

There were three major factors that became enablers to Japan's postwar recovery. First, the devastation of war created a sense of urgency. Second, Deming focused the Japanese on the pursuit of quality in manufacturing. Finally, the culture of the Japanese people made it possible to institutionalize quality into their manufacturing processes.

The innovations developed in Japanese factories spawned a worldwide epidemic of quality in manufacturing … and the rest is history. The fact that Japan became a formidable worldwide business competitor with the built-in advantage of greater experience in new manufacturing and management concepts severely tested the leadership of US industry. American factories using outmoded organizational structures, systems, processes, policies, procedures, and practices were forced to change the way they manufactured goods.

The efforts to improve US manufacturing over the years have focused on several areas, including the application of new technology, mechanization, process reengineering, just-in-time strategies, Lean manufacturing concepts, and the employment of Japanese-style management techniques. This has resulted in major improvements. But in spite of this, American managers don't have the luxury of becoming complacent. Even though the United States may have gained a little competitive breathing room, it is now facing even greater challenges.

Today, the competition isn't just Japan; it's the whole world. Aggressive manufacturing competition has evolved throughout the Asian/Pacific Basin, including South Korea, China, the Philippines, Singapore, Malaysia, and India.

In addition, the European Common Market is showing its competitive muscle, especially Germany. All of these countries are honing their own manufacturing skills, becoming better at what they do best, and in the process becoming a greater competitive challenge than ever before. And this will continue.

The future of US manufacturing depends on its ability to get more and more competitive and innovative. It cannot rest on its oars. Becoming more competitive is a never-ending exercise in mind conditioning for busy executives and senior managers. They recognize that their options for reducing costs are not unlimited. To continue wringing out savings in the places that have been visited again and again becomes more and more costly, with ever-diminishing returns.

Fortunately, American industry does have one area of unexplored opportunity that has never yet been leveraged: plant maintenance. Plant maintenance operations have the potential to become American industry's next competitive edge. The obvious question is, if this is true, why has it not been recognized?

Unfortunately, few (if any) plant executives or senior managers have ever been in a position to understand the extent that plant maintenance affects bottom-line profit. This is due in part to the traditional accounting practices used in most manufacturing companies, which do not provide a view of the total cost of maintenance. The largest costs of maintenance occur outside of the maintenance budget, in other areas of plant operation, as a consequence of maintenance requirements. In fact, it's common in US manufacturing plants for less than 30 percent of the total cost of maintenance to be reported in a manner that lends itself to controlling these costs.

While senior managers focus their attention on what they see as the controllable areas of high cost, they are not aware of the magnitude of controllable maintenance-related costs that impact bottom-line profits. This isn't surprising, because few if any senior managers have more than a casual knowledge about maintenance operations. Consequently, there is little understanding about how maintenance affects the accounting system outside of the maintenance budget.

Changes in manufacturing technology have brought about dramatically increased requirements for maintenance. Machine uptime and quality machine output requirements are now the highest in history. This, in turn, has created an ever-increasing dependency link between manufacturing capacity, production scheduling, and the effectiveness of plant maintenance. Just how well, or how poorly, plant maintenance responds to these situations will have a major impact on plant performance.

Technology is changing the way goods are manufactured. It can bring about major improvements in quality and reduce direct-labor requirements. But at the same time, it will increase the costs of maintenance, sometimes dramatically.

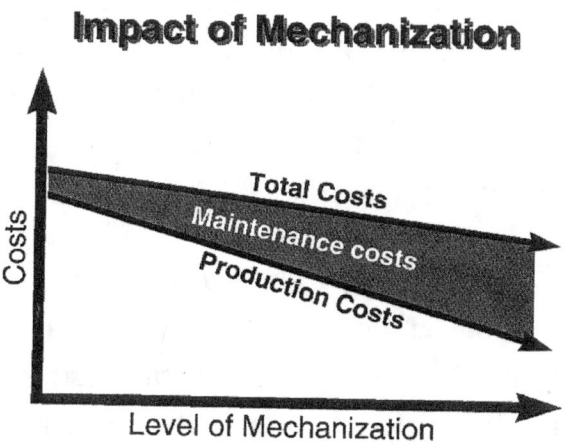

How much does maintenance really cost? Are maintenance costs too high? Are the level, quality, and consistency of maintenance services what they should be? Most companies can't answer these questions, either because they don't have the data or they simply haven't taken the time. If they could fully understand the economic implications of these questions, there would be a greater urgency to deal with these issues.

Certainly, maintenance costs are too high, and the level, quality, and consistency of maintenance services are rarely all they could be. However, don't assume that the maintenance organization is responsible, because this is almost never the case. Most often, the real culprits are dysfunctional systems and the nonsynchronous culture of the larger plant organization.

The financial view (or paradigm) of maintenance can keep people from seeing maintenance as a key enabler to the company's strategies for reducing costs. Plant maintenance has a direct effect on profit and loss because it is one of the few areas where a dollar saved contributes directly to profit. If manufacturing companies want to reduce costs, there are some very good reasons to focus on improving the effectiveness of their plant maintenance operations as a means to that end.

Chapter 2
The Paradigm Effect

Over the past forty to fifty years, there has been a strongly held management paradigm about plant maintenance, and there is an urgent need for a paradigm shift. It's important to understand the concept of paradigms. A paradigm is simply a strong, widely shared assumption: about government, politics, religion, scientific theories, education, anything! Paradigms can blind people to things lurking outside of the boundaries of their own paradigms—things that, if ignored, can have serious adverse consequences.

Joel Barker, in his book *Discovering the Future: The Business of Paradigms*, provides insight into this phenomenon. Barker cites other authors and various definitions of a paradigm as well as his own. He also defines the paradigm shifter (the person who creates a new paradigm), the paradigm pioneer (the early risk-taker who adapts to the new paradigm), the paradigm shift (the movement from the old paradigm to the new paradigm), and paradigm paralysis (the inability to anticipate or adapt to the new paradigm). Barker provides the example of a large consumer industry whose products once dominated the world marketplace but, due to a strongly held paradigm about those products, was unable to anticipate or adapt to a new paradigm that came into the marketplace and almost eliminated the need for its products.

Has something similar taken place right here in the United States? Detroit's big three dominated the world automobile marketplace for nearly a hundred years, but due to their strongly held paradigm about manufacturing automobiles, they didn't foresee that Japanese carmakers were creating a new paradigm that would rattle the very foundations of the US automotive industry. They seemingly didn't realize that the paradigm shift was already taking place within the automobile industry. Cars would become higher-quality and mostly compact and midsize, with better gas mileage, faster model changes, more innovations, and so forth.

Detroit's big three not only failed to anticipate the new paradigm, they also failed to respond to it quickly enough, giving the Japanese time to execute their entry into the US marketplace via exports to the east and west coasts, where they established their first dealerships before expanding later into the heartland of the United

States. In order to be successful, the Japanese not only had to provide products that combined quality, value, and economy—something that would appeal especially to younger first-time car buyers—they also had to counter the anticipated resistance of the United Auto Workers (UAW) and the ensuing "Buy American" campaign. Overcoming the perception that it was disloyal to buy a foreign car and possibly cause American autoworkers to lose their jobs was a formidable challenge.

The Japanese met that challenge by building their own manufacturing facilities here in the United States, hiring American workers to build Japanese-brand cars. A whole new perspective about buying foreign cars was established because relatives, friends, and neighbors were now building, buying, and driving these "Built in the USA" foreign cars. The Japanese have since become worldwide marketplace leaders and now are challenging Detroit's big three for marketplace leadership on their home turf, inside the country where the automobile industry was born. Is it possible that the big three of the future could be Toyota, Honda, and Subaru?

In spite of troubling times, there is a light at the end of the tunnel. The UAW have become stakeholders in both GM and Chrysler. A greater level of cooperation between the union and management increases the probability of success by orders of magnitude. It's no longer a matter of them versus us, because all that really matters now is what we can achieve together. And perhaps together, we can become the paradigm shifters of the future.

What has all of this to do with plant maintenance? The passing decades have provided a host of opportunities to see plant maintenance at its very best, and yet too often somewhat less. This isn't a criticism of the maintenance department or its staff, because in most cases they perform beyond expectations, considering that they so often work in an environment over which they have little if any influence or control. As a result, it's not surprising to still observe the same kinds operational dysfunction and consequences that existed forty to fifty years ago.

It's incredible in this day of modern management to see maintenance-related production costs and losses continuing to sap the economic strength of manufacturing companies; and worse, this seemingly being accepted as an unavoidable cost of doing business. It seems that management is unwittingly being blindsided by a closely held paradigm of plant maintenance operations. It is past time for a paradigm change.

Chapter 3
Maintenance 101

In order to understand the issues and opportunities associated with maintenance—and they are many—it's important to understand some of the most simple and basic facts about maintenance itself. Starting from this point of view, it doesn't really matter whether you are a maintenance professional, production manager, comptroller, or CEO. It's important for everyone to know the basics as a building block to appreciating how maintenance can contribute to the bottom-line financial objectives of a business enterprise.

Anyone having physical assets will have maintenance requirements sooner or later. It doesn't matter whether that asset is an airplane, home, punch press, automobile, or maybe just a lawnmower. While there is a cost to perform maintenance, there is almost certainly an even greater cost if proper maintenance is not performed. Failure to perform maintenance as required can result in temporary loss of use, costly repairs, or even the total loss of the asset itself.

The point is, there are maintenance costs associated with asset ownership, but proper maintenance can be highly cost-effective by minimizing the total cost of asset ownership. While the subject may become a little tedious at times, the information provided here is vitally important to understanding how all of this affects costs.

Some Basic Definitions

What does the word *maintenance* mean? Actually, it probably means a lot of different things to different people. The Merriam-Webster Dictionary provides the following definitions:

- **Maintain**: to cause to remain unaltered or unimpaired
- **Maintenance**: a maintaining or being maintained

A Google search provides the following definitions:

- **Maintain**: to keep in an existing state (as of repair, efficiency, or validity)
- **Maintenance**: the process of maintaining or preserving someone or something, or the state of being maintained.

The Google definition presents the term *maintenance* as a transitive verb, which it is. A transitive verb must have two or more objects. For example, the maintenance of *who* or *what*.

There are many different types or categories of maintenance work. For example, *aircraft maintenance* and *plant maintenance* are completely different activities. Both maintain physical assets; the one maintains aircraft and the other maintains machinery and equipment. Each type of maintenance has its own unique work processes.

Plant maintenance, sometimes called *factory maintenance*, is used here in the context of discrete manufacturing rather than process manufacturing, as in the case of an electrical-power-generating plant or a gasoline-refining plant. The objective here is to properly define plant maintenance and in doing so to avoid the vertical integration of plant maintenance into nonmaintenance activities. This continuing (and costly) practice has been taking place in many if not most US manufacturing plants over the past fifty to sixty years or more, resulting in immeasurably higher maintenance costs.

Any definition of plant maintenance must provide clear direction in determining *what is* and *what is not* plant maintenance work. In some manufacturing plants, there is a category of maintenance work that is classified as minor construction. As the word *minor* implies, it involves small jobs of short duration. The words *build, construct, install,* or *move* might be used in conjunction with a repair, periodic line changes, etc. Minor construction is not to be confused with costly large projects.

Unfortunately, in some manufacturing plants, major project and construction work is being performed by the maintenance department. This is done without an understanding of the costly negative effect on the core business of plant maintenance, which is maintaining the plant's physical assets. This costly, ill-advised practice, is discussed later in more detail.

I would like to propose the following definitions:

- **Maintain**: preserve physical plant assets
- **Maintenance**: the preservation of physical plant assets

If the work to be done doesn't match a definition such as this, then it's not plant-maintenance work.

Important Facts about Plant Maintenance

Plant maintenance is a specialized segment of the maintenance industry. Manufacturing plants traditionally have maintenance work performed by their own maintenance department. This department is staffed by highly skilled technical workers.

Plant maintenance personnel often maintain hundreds or even thousands of individual physical assets. Such work commonly involves two types of activity: *production maintenance* (maintaining production machinery and equipment) and *facility maintenance* (maintaining buildings and grounds, HVAC, utilities, process supply systems, waste treatment, etc.). Except for the scheduling of routine maintenance, the work is primarily nonrepetitive, with each job being unique and creating new learning challenges.

There is usually a central maintenance shop that has special machinery and equipment for performing certain types of work. Work performed outside of the central shop may be anywhere inside, outside, or even on top of the plant's buildings. Maintenance skilled-trade employees are among the highest paid hourly workers in the plant.

Plant maintenance is traditionally funded with an annual budget. It is usually the highest-cost indirect expense item in the plant's accounting system. Plant maintenance costs increase with the increased use of complex high-tech manufacturing machinery and equipment.

Maintenance-related costs occur outside of the maintenance department's budget in other areas of plant operation. These costs are commonly much greater than the maintenance department budget. Production has become increasingly dependent upon plant maintenance workers to achieve quality, cost, and delivery objectives.

The Role of Maintenance Management

Maintenance management involves the following:

- defining mission goals and strategies
- developing operating policies, procedures, and practices
- identifying measures of business performance
- structuring and staffing the organization
- establishing and managing operating budgets
- providing machinery, equipment, tools, and workspace
- providing logistics support
- determining scope and levels of maintenance service
- establishing work methods and practices
- training and developing people
- fostering and facilitating the desired organizational culture
- establishing effective levels of communication
- identifying data resource requirements
- institutionalizing measurement and control
- authorizing and prioritizing requests for work
- planning and scheduling work
- allocate and assigning resources
- monitoring work progress, quality, cost, and completion
- ensuring compliance with health, safety, and environmental regulations
- ensuring compliance with legal and contractual requirements
- facilitating continuous improvement
- recognizing and rewarding contributors to success

People and *quality* are two factors that require special consideration in plant maintenance operations. In fact, they are major factors in determining the success or failure of any organization. One important people-related study, conducted in 1927, looked at human behavior in the workplace. This study was conducted in Chicago at the Hawthorne Plant of Western Electric. It has since been called "The Hawthorne Studies."

In these studies, researchers separated out a small group of workers to determine the effects of making different kinds of changes in the workplace. At first, they introduced a series of improvements that resulted in increased productivity. Then they took away some of those improvements; surprisingly, productivity increased again. Then they began a series of alternating improvements and then taking them away. The results were always the same: productivity increased.

The conclusion of the Hawthorne Studies was that "When workers feel that management is paying attention to them, they respond in a positive way." Pay attention to the worker, and the worker will take care of the job. Although the Hawthorne Studies involved only hourly workers, it's probable that the same holds true for everyone, at every level.

Quality Maintenance

The importance of quality cannot be overemphasized. The late quality guru Philip Crosby would have described plant maintenance as a quality process, and it is. The following attributes of quality defined by Mr. Crosby's quality management concepts can be applied to plant maintenance:

- Quality is conformance to machine uptime requirements.
- Prevention is the process for achieving quality.
- The quality standard is zero downtime.
- Quality is measured by the costs of machine downtime.

Machine downtime is another area where far greater understanding is needed. The production costs and losses associated with machine downtime go beyond maintenance department costs. The production costs and losses are appropriately called *maintenance-related costs*. Maintenance department costs plus maintenance-related costs make up the total cost of maintenance.

Costs of Machine Downtime

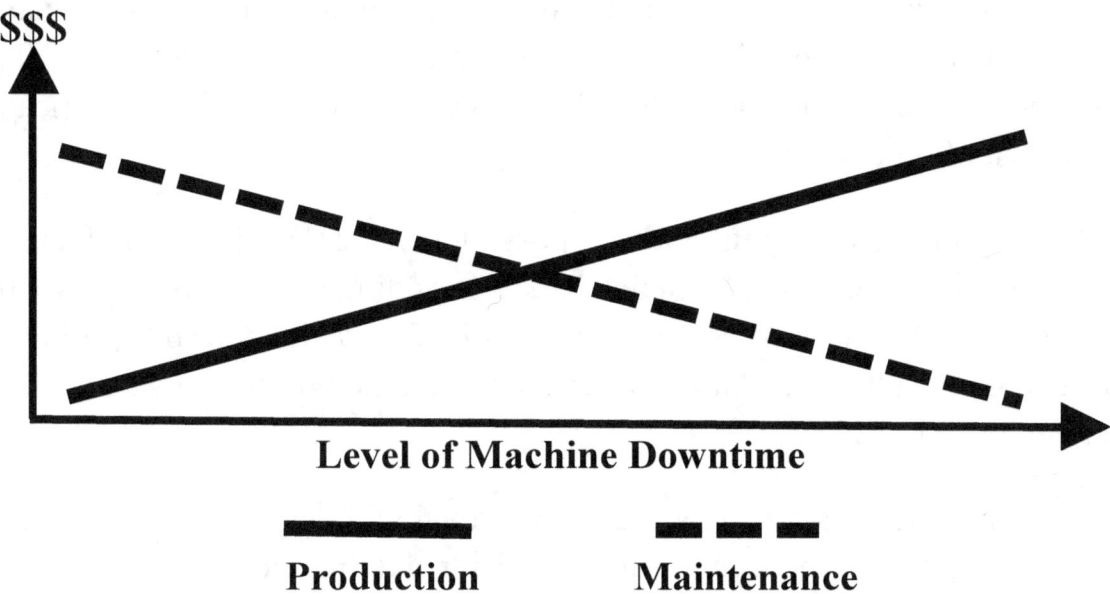

The graph above demonstrates the cost relationship between production and maintenance when the maintenance focus is on reactive-corrective actions. Note the point of equal minimum total costs. The graph below demonstrates the same cost relationship between production and the maintenance department when the maintenance focus is on prevention rather than reactive corrective actions. Note that the point of equal minimum total costs and the level of machine downtime are all lower.

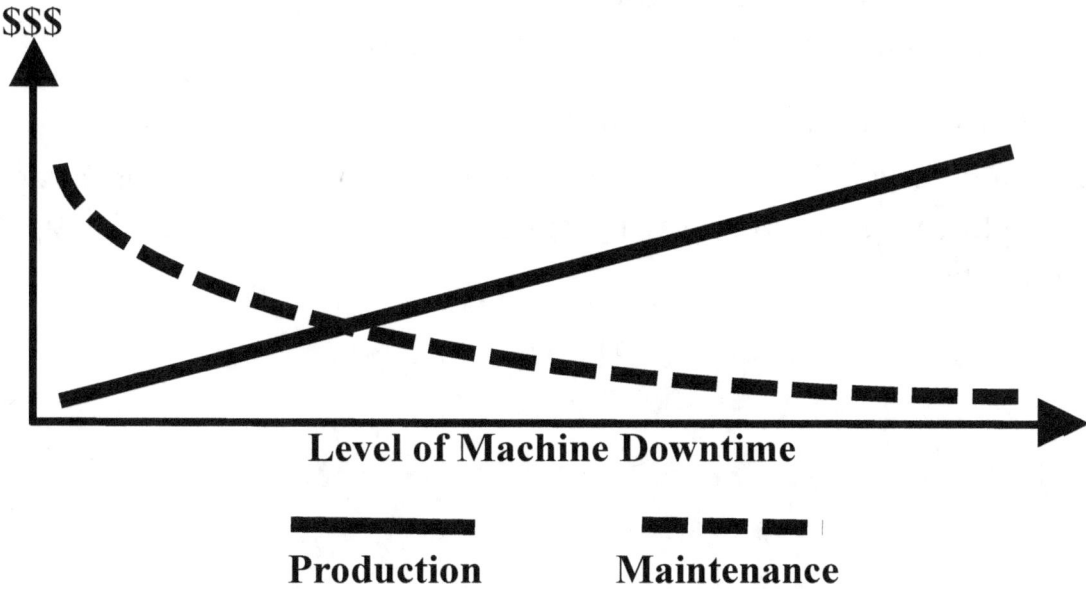

These simple but important facts lay the foundation—a first step in the process of understanding how plant maintenance can become American industry's next competitive edge. It's my conclusion that in many, if not most, large manufacturing plants, major on-the-job work constraints are adding significantly to the cost of production.

Reactive maintenance must (where possible) be replaced with a focus on preventing the requirement for maintenance. In addition, the focus on prevention must go beyond the traditional concepts of SPM to include preventing the constraints, roadblocks, barriers, and delays that are so prevalent in the maintenance workplace.

Maintenance workers are among the most loyal and dedicated workers in the plant, but their job is made much more difficult and time-consuming by on-the-job constraints they face on a daily basis. This situation translates into increased levels of machine downtime along with its associated production costs and losses.

There are four specific improvement strategies that can be implemented in a relatively short period of time, and they can bring about major cost avoidance and cost savings:

1. Real-time mobile-communications for maintenance
2. Centralized dispatching of maintenance resources
3. Flexible zone configuration for production support
4. Condition-based production priority system

These four areas for bringing about improvement (and others) are discussed later in more detail. The reason for mentioning them here is to emphasize up front that there are major cost savings that can be achieved in a relatively short period of time.

Chapter 4
Dollars and Sense

To those engaged in discrete manufacturing and struggling to compete in the face of aggressive worldwide competition, take heart: there's some good news. American manufacturing companies almost certainly have the means within their own organizations to create a new competitive edge. The bad news is that they may not recognize the opportunities that exist because of their own closely held business paradigms. After all, how many key plant executives have ever considered a strategic initiative focused on plant maintenance as the means of improving plant performance?

In 1972, *Dunn's Review* reported that maintenance costs in the United States were between $100-$200 billion annually. By 1990, it was estimated that maintenance would cost approximately $600 billion annually and continue to rise. In addition, up to a third of all money spent for maintenance may be wasted. From a plant perspective, maintenance expense is commonly 15 to 40 percent of the cost of production; the average is 28 percent. Maintenance is usually a plant's largest indirect expense. It's also the fastest growing segment of cost, increasing 10 to 15 percent per year (The above data was published in P/PM Technology MMS Report in 1980).

What makes this situation worse is that these numbers represent only a fraction of the actual cost of maintenance. They're just the tip of the iceberg. It was estimated by one of the big three US automobile manufacturing companies that it cost them approximately five dollars in leveraged costs for every one dollar spent on maintenance labor and material. (I will substitute the term *maintenance-related costs* for *leveraged costs*; it seems to be more descriptive of the costs in question. These maintenance-related costs appear elsewhere in a plant's accounting system.)

This estimate came as no surprise to maintenance professionals who have long understood that the total cost of maintenance was far greater than the plant's maintenance labor and material budget. This isn't a single-company problem; it's an industry-wide problem. Virtually all discrete manufacturing plants are faced with the same dilemma of enormous total maintenance costs.

While the maintenance department budget is significant, the maintenance-related costs that occur outside of the maintenance department budget, in other areas of the plant operation, as a result of machinery malfunction and failure are significantly higher. A strategic focus on plant maintenance may not seem logical to everyone, but that is exactly what I am suggesting. There is ample evidence to support the idea that plant maintenance represents one of the last great areas of unexplored opportunity for American manufacturing companies to become more competitive in the worldwide marketplace.

The following are some of the most common maintenance-related costs and losses:

- **Direct labor losses, inefficiency, and overtime**—When a production machine is down due to a failure, the operator is idle until the impact of the failure is known. If it's determined that the machine will be down for only a short period of time, the operator likely will remain idle until the machine is restored. If a longer downtime period is indicated, the operator may or may not be moved temporarily to perform other work.

- **Lost production due to machine-related downtime**—When a machine stops due to a failure, production stops until the machine is restored or the job is transferred to another machine. These delays are costly. When work resumes, the lost production can be made up by overtime or scheduling additional shift work. This is also costly.

- **Premium freight**—Machine downtime results in expensive premium shipping in order to avoid or minimize late customer deliveries.

- **Relocation and setup of machine tooling**—Machine downtime often makes it necessary to move the tooling from the down machine to another machine in order to restore production as quickly as possible. These delays and added work add cost.

- **Reworking and repair of production parts**—Malfunctioning machines cause deviations that require costly rework or repair of production parts.

- **Creation of product scrap**—Malfunctioning machines cause costly production parts to be scrapped.

- **Product quality and reliability problems**—Malfunctioning machines create product reliability problems, which emerge only after the product is shipped to the customer, resulting in added warranty costs.

- **Idle in-process banked-part inventories**—In order to avoid interruptions to production due to machine downtime, it has been common practice to bank some level of in-process production parts at certain workstations. In the event of machine downtime, these part inventories provide a temporary supply of parts to downstream operations. Although these banked part inventories may reduce the risk of downtime-related production losses, they also increase the cost of doing business by having costly idle inventory.

- **Wasted energy and other plant resources**—Looseness in bearings, seals, gaskets, and joints results in the continuous loss of lubricants, hydraulic oil, compressed air, steam, water, and gases. Leaking or sticking shutoff valves and steam traps add to these losses. Also, malfunctioning control systems waste energy. The list goes on and on. These and other examples of costly waste are all too common in many manufacturing plants.

- **Premature wear-out and replacement of machinery and equipment**—When the level, quality, and consistency of routine maintenance is less than it should be, or when there is improper application or operation of machinery and equipment (such as excessive speeds and feeds, improper adjustments, ignoring incipient machine problems, operator error, etc.), premature wear-out and failure can occur. Overhaul and rebuild are also expensive.

- **Backup machinery and equipment**—In plants experiencing high levels of machine downtime, the procurement of additional machinery and equipment to ensure production capacity is common. This is a very expensive solution; effective preventive maintenance is far less expensive.

These are all a part of maintenance-related costs and losses. They are not imagined or made up. They are painfully real, and they far exceed the cost of maintenance labor and material.

According to Maintenance Consultant, Terry Wireman, in his book "Total Productive Maintenance" published in 1991, he states " In addition to these costs, it's common that 30 percent of all spending for maintenance is wasted."

If and when the economic impact of these maintenance-related costs and losses are made visible to upper management, and management understands that a large majority of these costs and losses are preventable, then managing the total cost of maintenance will become more of a strategic issue.

While some may look at this situation as a big problem, hopefully there are others who will see it as a big opportunity for improvement—and they will go for it.

Chapter 5
Managing by the Numbers

It seems unreasonable that the maintenance department doesn't have the tools to go beyond just fixing things, because maintenance workers could be so much more than just an expense. They could be making a significant positive contribution to the bottom line of the business enterprise—if only they had the numbers to manage by.

There is an old adage that "You cannot manage what you cannot measure." Without good measurement (data), it's not possible to effectively control maintenance costs in a manufacturing plant. In most cases, the only thing being measured is the maintenance budget, and that's part of the problem. Any discussion about the maintenance budget should be viewed in the context of larger maintenance-related costs and losses occurring outside of the maintenance budget. If the maintenance department is to be responsible for reducing both its own costs and costs that occur elsewhere in the plant due to maintenance, more data will be required.

There are some important questions that need to be asked in this regard. Is the current maintenance budget too much or too little? What criteria are being used to determine the size of the maintenance budget? Are questions such as these being asked during the budget-making process? If not, why not? Someone should be asking a lot more questions—and expecting answers.

There are serious questions to be asked about the maintenance budget. First and foremost, the maintenance budget is an investment. Is there any other investment the size of the annual maintenance expense budget that would ever be approved without requiring an acceptable return on investment (ROI)? If that question suggests that the maintenance budget approval process should require some measurable return on that investment … absolutely!

The objective of the maintenance department is to minimize the total cost of maintenance for the plant, which includes production costs and losses due to maintenance-related machine downtime. Establishing specific reduction goals in maintenance-related costs and losses should be a required part of the maintenance budgeting process. That's where the ROI comes in.

In many plants, the budget process uses the current budget year actual expense as the basis for forecasting the budget for the following year, considering adjustments for production volume projections and other factors. This is done without having to justify the current budget, which can't be done using existing maintenance accounting methods. At the present time, it's common for maintenance accounts to be nothing more than big bags with titles such as labor, tools, parts, materials and supplies, and training.

It's the use of these big accounting bags that makes it virtually impossible to justify the existing budget or the need for changes. How and where maintenance resources are being utilized can only be generalized using these big bags.

On the other hand, activity-based costing (ABC) breaks down spending into smaller accounts that identify more exactly how and where maintenance spending takes place. ABC provides documentation that makes the budgeting process easier by orders of magnitude. For example, based on the summary of individual job assignments by employee classification data (such as electrician, machine repair, pipefitter) for the budget year, information could easily be reported in man-hours by job classification, machine, department, and plant.

ABC can also identify the type of work being performed, such as scheduled routine maintenance, unscheduled repairs, preventive maintenance, HVAC, line changes, grounds maintenance, relamping, and rebuilding spare parts. Parts, materials, and supplies could be charged and reported by the physical asset number. This would give management a first-time-ever overview of how and where maintenance assets are actually being utilized.

In addition, ABC could provide comprehensive machine history data; identify and compile repair actions taken and parts and materials used; and track repetitive failures. This data would directly support various preventive activities for improving machine reliability and maintainability—the objective being to reduce machine failures, production downtime, and related production costs and losses. This would not only provide upper management with its first look at precisely how maintenance dollars are being spent; it would also point to a quick and easy way to budget for the future.

There are many ways ABC data can be used. For example, if additional machinery and equipment is being added, it provides a data-based method of forecasting requirements for the new machine. The same is true when disposing of machinery and equipment.

Managing the total cost of maintenance requires data-based tools that few, if any, maintenance departments have at their disposal. The reason is that much of that data is buried somewhere within the plant's general accounting system (GAS). Some data about maintenance-related production costs and losses probably does exist in the GAS.

Tracking and managing the total cost of maintenance isn't the same as balancing your checkbook, because the money you are spending takes the form of continuous workplace activities taking place on the plant floor to maintain a plant's physical assets. This is where you spend the money, and this is where the opportunities are for saving money.

The illustration below depicts the current sad state of maintenance management in many if not most manufacturing plants.

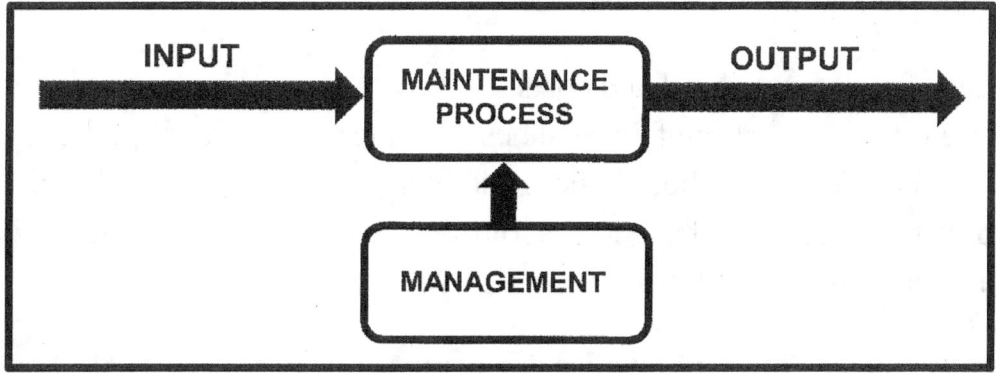

Unfortunately, as pictured above, the maintenance manager inputs information to control the process but gets no feedback (data) to know whether the actions taken were effective. It's worth repeating over and over again: "You cannot manage what you cannot measure." On the other hand, if you have a true management system, as pictured below, the maintenance department would have the tools and accountability for achieving the minimum total cost of maintenance.

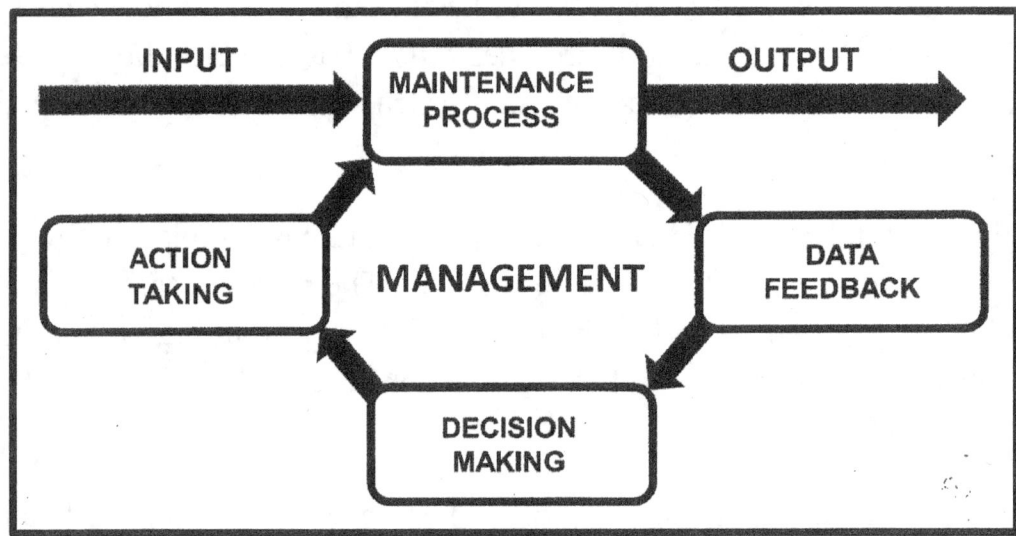

The good news is that most modern computer-based accounting systems allow for the use of special coding that can extract many of these maintenance-related costs for exception reporting without interfering with the normal routine accounting process. If it's not possible or practical to isolate some of these costs, engineering estimates can be used. The actual method of measuring these costs isn't as important as ensuring relevance and consistency; nor is the nth degree of accuracy required in every instance.

Examples of maintenance-related costs are clearly seen when there is production downtime due to a machine failure. These maintenance-related production cost and losses include direct labor idleness and inefficiency, overtime to make up for lost production, premium freight to avoid late deliveries, reworking production parts, and parts being scrapped.

This doesn't include the maintenance department costs (labor and MRO parts and materials to restore the machine to production). Many other forms of maintenance-related costs are not quite so obvious. They are buried within the GAS.

The objective is to track relevant categories of cost and provide actionable management reports that can be used by maintenance to monitor the outcome of specific actions taken. This step-by-step process of measuring the effects of actions taken is the road to continuous improvement. This is managing by the numbers.

Chapter 6
Constraints, Barriers, Roadblocks and Delays

Even though the objective of SPM has always been to prevent machinery and equipment failures, prevention of *all* failures has never been achieved and probably never will be. Nevertheless, there are prevention strategies that can be put into place before failures occur to significantly reduce the adverse effects of failure.

It has been clearly demonstrated over time that when machine failures occur, they are very often accompanied by constraints that delay the restoration of the failed machine, adding to both the cost of repairs and the associated downtime-related production costs and losses. The majority of these are preventable. If the unnecessary added cost of constraints and delays are to be avoided, it's essential for senior managers to be aware of how these costs and losses are incurred—and how they can be prevented.

It may not be easy for management to fully accept what can and too often does take place, beginning from the moment of failure until the machine is restored to production. This is demonstrated by mapping a chronology of events when a production machine fails. Remember, not every failure is a quits-running situation. A failure is defined as any machine condition that has a negative effect on product quality, cost, or delivery.

Most machine failures require assistance from plant maintenance to make the necessary repairs. A machine failure triggers a series of events, listed below in the order in which they would normally occur:

1. Failure
2. Failure detected
3. Repair assistance requested
4. Technician assigned to repair job
5. Repair performed
6. Machine released to production

A firsthand look at how the chain of events too often unfolds during a machine failure may read like a Keystone Cops comedy, but don't be misled. These are common everyday situations associated with machine failures, and they are far too costly to be the least bit amusing. It's within and between these events that most of the costly work constraints and delays occur. These dysfunctional scenarios are called *constraint events*.

Sometimes a failure can happen very suddenly, and the machine stops running. At other times, failure will evolve over a period of time; the machine keeps running while the integrity of a part or component degrades, and in time it will fail. These evolving failures are the most preventable of all.

Scheduled routine maintenance inspections (if performed correctly) will detect these problems in the incipient stages. Then corrective actions can be taken before a more costly constraint event occurs. This is the essence of prevention.

The first constraint event often takes place after a machine failure occurs but before it is detected. The machine operator may not be immediately aware that a machine failure has occurred because the point of failure is not visible to the operator. For example, when an automatic parts loader becomes jammed, the machine continues to cycle, and the operator is unaware of the problem until the absence of parts reaches the operator's workstation. Consequently, both production and direct labor are lost. Similar failures are quite common.

Machine failures become more costly when they result in production part defects. If the operator isn't aware of the problem and the flow of defective parts continues until found at a later point in the manufacturing process, this can result in costly in-process repairs. If the defect isn't found until the product has been fully assembled, that product may have to be scrapped, which is even more costly. Of course, the worst possible scenario occurs when the part defect isn't found at all during manufacturing and the finished product is shipped to the customer, who later experiences quality and reliability problems. This can become more than just a warranty cost issue; it may jeopardize future sales to that customer.

Another example of preventable added cost takes place when an undetected machine problem is evolving and the machine continues to run until there is a sudden failure. Sometimes this can result in major damage to the machine, which would mean extended machine downtime, costly repairs, and associated downtime-related production costs and losses.

Once the failure occurs and is detected, another preventable delay may take place. While the operator is trying to figure out what happened to the machine, the clock is ticking, and nobody has yet called the maintenance department for assistance. The question is, who is supposed to call the maintenance department: the machine operator, the utility or setup person, or the foreman?

If this situation seems unlikely, it really isn't. Employees move from department to department and from shift to shift over time. Foremen also move around, and unless established procedures are in place before a failure occurs and all workers (including those new to the job) are properly trained and understand whose job it is to call for maintenance assistance, these kinds of delays actually do take place.

Suppose in the case cited above, the machine operator notifies the production foreman, who calls the maintenance department—but the line is busy, and the foreman must make several attempts before getting through. When contact is finally made with the maintenance department, the foreman isn't there; he's out in the plant somewhere. He is then paged on the plant's paging system. When there is no response after a few minutes, the page is repeated and still no response; the maintenance foreman may be in a noisy location or on the roof, where it's almost impossible to hear the paging system.

Upon returning to the maintenance office, the foreman learns of the attempts to contact him, and he immediately calls the production department and is informed about the machine problem. Since there have been repeated mechanical problems with this machine, the foreman goes to the maintenance shop to locate Tom, a machine repairman who has worked on that machine recently. However, Tom isn't there, so he is paged. Several minutes pass before Tom calls back and is assigned to the job.

Actually, this is a reassignment for Tom, because he had already been assigned to repair a power-drive transmission, which in this case is of lesser importance than getting the failed machine running. This requires him to extract himself from the job he had been working on. Tom has to gather all of his tools together, secure the area for safety purposes, and then travel to the site of the failed machine. This adds another fifteen minutes of downtime for the failed machine.

When Tom finally arrives at the machine, it's several minutes before he can locate Paul, the machine operator, to discuss the problem. After several unsuccessful attempts to cycle the machine, it's determined that the problem is probably electrical and not mechanical. Paul then calls the maintenance office to inform his foreman of the need for an electrician and learns that the foreman is out in the plant somewhere. He pages the foreman, and it's several minutes before the foreman calls back and learns that an electrician is needed; however, no electrician is available, as they are all currently working on other high-priority jobs.

These kinds of delays are the result of multiple simultaneous machine failures, and they occur quite frequently in medium and large manufacturing plants. The wait for an available electrician may be short, but unfortunately, it could be a long wait.

Eventually Fred, an electrician, is assigned to the failed machine. After making a brief inspection, he travels to the maintenance office to obtain the machine manual and wiring schematics. He spends considerable time looking for the manual because it isn't where it's supposed to be—a common problem. Since the master copy is located in the engineering library, Fred travels to the engineering office and waits while selected pages of the machine manual and wiring schematics are copied for him. Having obtained those items, he travels to the maintenance crib to check out a digital oscilloscope, which will be used in diagnosing the electrical problem.

With everything needed finally at the jobsite, Fred begins the process of diagnosing the problem. Eventually, he discovers that the controls are wired differently from the electrical schematics he is using. He then contacts his foreman and requests engineering assistance. Thirty minutes later, an engineer arrives at the jobsite, and Fred explains the issue of the controls not being wired according to the electrical schematics. The engineer proceeds to check the drawings with the control wiring and confirms Fred's conclusion: the wiring and the print are different. Upon returning to the engineering library, Fred finds a set of hand-drawn electrical revisions that had never been updated in the master file (another all-too-common problem).

Fred then returns to the jobsite with a copy of the electrical revisions and soon finds the culprit: a problem in one of the logic boards. When he arrives at the parts crib, he is told the logic board is out of stock. Since it's Friday, a replacement can't be expected before Monday or later, even if it is expedited.

Fred again calls the maintenance office, and fortunately, his foreman is there. He explains the problem and suggests temporarily using a logic board from A-12 in department 88, which uses the same logic board. Together, they proceed to the production department to talk with the foreman there about the possibility of temporarily borrowing the control logic board from A-12; that is, if the machine isn't scheduled for production for at least three to four days.

The production foreman is reluctant to agree, stating, "What if one of my machines goes down? I wouldn't have a backup. We had better discuss this with my superintendent. If it's okay with him, then it's okay with me."

The three of them proceed to the superintendent's office, where they learn that he is in a meeting and isn't expected to return for at least another thirty minutes. Fred is told to get together whatever tools he would need and to wait at the maintenance office until the issue is resolved. Twenty-five minutes after he arrives at the maintenance office, his foreman returns with permission to borrow the logic board. Fred tells his foreman that due to the configuration of the controls on A-12, he will need a little temporary help in accessing and removing the card; his foreman agrees to send someone to help him. Fred proceeds to A-12 and works to the point where extra help is needed—but before help arrives, it's time for lunch.

During lunch, Fred reminds his foreman that he is waiting for assistance. He's told that someone will be there immediately after lunch. When Fred arrives back at A-12, it's another fifteen minutes before someone arrives to help him. Thirty minutes later, the logic board is removed, and Fred is back at the site of the failed machine. He proceeds to install and test the borrowed part.

When Fred is ready to check out the operation of the machine, the operator is nowhere to be found, so Fred informs the production foreman—only to learn that Paul, the operator, has been temporarily assigned elsewhere as a float operator. The foreman agrees to send him back just as quickly as he can find a replacement. Thirty minutes later, Paul arrives, cycles the machine, and indicates that everything appears to be working correctly. The machine is then released to production. Tom continues to monitor the machine for a few minutes before returning to the maintenance shop.

Does this scenario seem rather bizarre, even ridiculous, a figment of the imagination? While it might be difficult for some managers to accept the fact that these kinds of preventable delays are taking place in their plants, they almost certainly do. Obviously, all of these things don't happen on every job; but some of these costly delays will take place on most jobs. Many of these are seen only in bits and pieces as incidents quickly forgotten in the face of the continuing challenges of getting the current job done. Workers don't have the luxury or the time to stop and reflect on what is really taking place, and far less time to actually do something about it, even if they could.

During my many years in the plant maintenance environment, I have personally witnessed the kinds of delays cited here, plus many, many others. The culprit is not the maintenance department, which makes extraordinary contributions to the plant in spite of a host of obstacles and constraints over which it has little or no control. Maintenance people demonstrate their professionalism, loyalty, and value to the business enterprise in innumerable ways, day after day after day. Unquestionably, however, a major problem exists, and it's essential that a plant's decision-makers understand the root causes.

These are issues for senior managers. They, and they alone, are in a position to make the decisions and take the actions needed to change the culture of waste that has existed for far too long in US manufacturing plants. The needed changes won't be accomplished overnight or without some stress and pain; but change can bring about major improvements that will be reflected in the bottom-line profits of the business enterprise while removing a lot of the job-related frustrations workers face every day, making them more productive and bringing them greater satisfaction in the workplace.

The good news is that the majority of these systemic constraints can be reduced or eliminated, more often than not, by changing management controls, such as organizational structure, systems, processes, policies, procedures, and practices. Remember, this requires a global, plant-wide approach rather than a focus on business segments, departments, or individual workgroups. The issues involved transcend traditional boundaries; that's why the active involvement of senior management is so necessary. Nobody else is in a position to make the necessary changes. The ball is in their court.

Chapter 7
Preventive Maintenance
(The Crown Jewel of Plant Maintenance)

If preventive maintenance (PM) is so great, why do so many people feel unhappy about their own plant's PM program? Could it be that there isn't enough time and resources? Could it be that there isn't enough management support? Could it be that nobody has taken the time to put together a cost-benefits proposal?

Whatever the reason an effective preventive maintenance process is not in place, the people who should be complaining the loudest are those in production management, because the ultimate benefit from PM is reducing the cost of production. It has been estimated that maintenance costs average 28 percent of the cost of production. Production should be the driving force behind PM. Production should be insisting on it.

What Is Preventive Maintenance?

Simply stated, PM is a set of structured processes for reducing or eliminating the requirement for maintenance. There are two major types of PM: nonrepetitive and repetitive. Nonrepetitive PM is performed once, and the benefits continue throughout the life cycle of the machine. This is the most cost-effective form of PM.

Consider that when new machinery and equipment are purchased, maintenance will be the largest future cost of the asset, continuing over its entire life cycle. Maintenance costs may ultimately exceed the original cost of the asset itself. When new machinery and equipment are to be purchased, this is when and where preventive maintenance can save a lot of money.

Prior to the procurement of a new machine, having both maintenance and engineering review the subject machine, looking for opportunities to improve its reliability and maintainability, can be an invaluable first step in PM. A special focus is placed on wear parts, components, and controls, comparing them to what experience has shown to be the highest quality available. Based on the outcome of this inspection, reliability and maintainability specifications can be included in the procurement process. Obviously, this may increase the initial cost somewhat. However, any increase in cost should be weighed against the estimated life-cycle cost of maintenance.

What about plants that may not be purchasing new capital assets? No problem; there will be a host of opportunities within the plant's existing machinery and equipment. Every plant has its own problem machines. Look for ones with a history of repetitive problems, especially among the plant's most critical machines. This is where PM will have the greatest impact. Have maintenance and engineering jointly review the history of recurring problems and put together a proposal to modify those machines to improve their reliability and maintainability.

That all sounds so easy to do. If that's all there is to it, why isn't everyone already doing it? There is a simple answer for that: it's because of where the focus is. Maintenance is so busy fixing things, and they have become so good over time doing that, it has become their modus operandi. The objective is to change the focus from fixing to preventing.

Upper-management action may be required to overcome the inertia involved. A directive to manufacturing, engineering, and maintenance to identify machinery and equipment where modifications would improve machine reliability and maintainability could start the process. Choosing the right machines will make its own business case. Remember, the benefits of modifications continue for the life cycle of the machine. These include reduced machine downtime, reduced maintenance requirements, reduced production costs and losses, and reduced overall cost of production. It usually takes just one good success story to energize the focus on prevention.

The second type of PM, which is repetitive, must be continued throughout the life-cycle of the machine in order to continue its benefits. This is the most common type of PM in most manufacturing plants. Repetitive PM requires careful management of the process to insure it doesn't cost more money than it's saving. More on this later on.

CLAIR

The widespread use of repetitive PM has its roots in the maintenance of military equipment during World War II. In conjunction with this, something called the CLAIR activities has emerged:

- *C*leaning
- *L*ubrication
- *A*djustment
- *I*nspection
- *R*eplacement

These five PM activities originally became the basis for what is known as *scheduled routine maintenance*. Just for a point of clarification, all repetitive PM should become scheduled routine maintenance.

Cleaning

There may be some who don't consider cleaning to be a PM activity. Try telling that to the Japanese. Many Americans who have visited Japanese factories at first glance thought all of the machinery and equipment was new, only to find out that what looked like new could have been forty to fifty years old. The machines looked new because the Japanese kept them clean.

There is no question that dirt, grit, and grime will deteriorate a machine's condition over a period of time. At the same time, clean machines contribute to happier, more productive employees. Keeping machinery and equipment clean is an important preventive activity.

Lubrication

If there is any one PM activity that's more important than others, it's probably lubrication, because of the major damage that can occur if it's not done properly. It can't be too little, or too much, or too late; it needs to be done right every time.

It's a sad state of affairs when something as important as lubrication is assigned to someone classified as an oiler—usually one of the lowest-paid jobs in maintenance. The lubrication technician deserves professional training, recognition, and pay for the all-important job of lubricating a plant's machinery and equipment. The role of the lubrication technician can be expanded to include activities like oil analysis and other associated activities.

Adjustment

Machines don't usually keep themselves permanently in proper adjustment. There are many things that can get out of adjustment, including pressure, temperature, speed, humidity, voltage, and travel. Keeping things properly adjusted not only prevents machine malfunction and failure, it can also prevent damage to production parts.

Inspection

Scheduled routine machinery and equipment inspections can prevent machine malfunction and failure by detecting incipient problems, which can then be corrected before a failure occurs. This is the essence of prevention.

Replacement

Routinely replacing machine parts or components prior to the end of their useful life cycle prevents machine failure, downtime, and associated production costs and losses.

It has been from the initial application of CLAIR activities that preventive maintenance as we know it today has evolved.

Getting Started with Preventive Maintenance

Where do you start the in-house PM process? Focus first on critical production machinery and equipment, plus large motors, generators, air compressors, process supply systems, and other powerhouse applications. After that, remember the 80/20 rule, because it probably applies here: 20 percent of the machines cause 80 percent of the problems. Start with the 20 percent.

Use the machinery and equipment owner's manuals as the primary resource for determining what PM tasks to perform. Anything that experience has shown the need for can be added. Be conservative; don't add too much too quickly. Make sure the process is working correctly before expanding the effort.

PM should be worked into production scheduling. All PM that can be performed during operation of the machines, or during a brief shut-down, should be done at that time. It's important where possible for production to actually see the PM work being performed.

In addition, there needs to be documentation at the machine that the PM work has been completed. A PM log at the machine should be provided for employees to input two important pieces of information: the date PM was completed and the initials of the person who performed the work. That will become really important data if there should be some kind of unfortunate accident and OSHA comes calling.

Employees who regularly perform PM often work on an off-shift. Nobody sees what they do, and they never get to see the results of their efforts. It takes a special kind of dedicated employee to work in that environment day after day, making sure that everything they do is done right and on time. There should be special recognition for the people who can do that.

Random audits should be conducted to ensure that scheduled PM tasks are being completed. This is necessary because experience across multiple plants has shown that it's not uncommon for PM tasks to be reported as being completed when the work was never actually performed. Unfortunately, this is probably more common than people realize. Failure to perform PM when needed can have costly consequences. It's best to see some evidence that PM work is being completed. The fact that audits are performed can be a deterrent to false reporting.

One way to help ensure that PM is being performed correctly is to provide instructions with enough detail that someone who has never performed the task before can do it correctly the first time. In addition, the task must include a list of any special tools, equipment, and supplies needed to perform the task. The following are two examples of a very simple lubrication task. The first, unfortunately, is all too common:

> ID: 342118
>
> Description: No. 16 Injection Molding Machine
>
> Location: Plant 1, Department 95, G-C4
>
> Frequency: Monthly
>
> ---
>
> Task:
>
> Lubricate hydraulic pump motors

The second example below provides enough information for someone performing the task for the first time to do it correctly.

> ID: 342118
>
> Description: No. 16 Injection Molding Machine
>
> Location: Plant 1, Department 95, G-C4
>
> Frequency: Monthly
>
> Task:
>
> Lubricate (2) hydraulic pump motors by removing both the filler and relief plugs and adding grease. Use 'Citgo HEP-2' or equivalent. Before adding grease, be sure the relief port is open to allow the free flow of excess grease.
>
> <u>Failure to do this will cause the seals to be pushed out of the bearings.</u>
>
> Run motors until the excess grease has vented, then replace the plugs.

Having fully descriptive PM tasks is an essential element of quality preventive maintenance.

Using the information provided by the owner's manual, plant engineering, and maintenance people a master schedule for all of the plant's physical assets can be developed. It's recommended that the equipment ID number and a plastic-encased copy of the PM tasks to be performed be attached to the equipment itself in a highly visible location. In addition, PM scheduled activities should be routed whenever possible (from machine to machine) to minimize travel and maximize labor efficiency.

There are numerous low-cost commercially available PC programs that can be used for scheduling PM activities. These programs use common time-cycle scheduling methods, such as:

- weekly
- semimonthly
- monthly
- bimonthly
- quarterly
- semi-annually
- annually
- user-specified

Trends in PM

Preventive maintenance didn't change much until the 1980s, when *predictive maintenance* was introduced to the world. What is predictive maintenance? It's simply the use of modern technology in the form of precision instruments to inspect machinery and equipment—the objective also being to determine when maintenance should be performed, rather than performing maintenance based on a time-scheduled basis.

The most publicized, advertised, and discussed PM process has been vibration analysis. Manufacturers, vendors, trade publications, and the consulting industry pulled out all of the stops for its introduction. It was presented as if it would become the Holy Grail of plant maintenance activities.

There is no question that vibration analysis, along with other technologies, can be important PM tools. However, vibration analysis is expensive to implement. In addition, later technologies, such as infrared monitoring, have proved to be more cost-effective and have a wider range of application, including the detection of both mechanical and electrical abnormalities.

Another approach to consider is outside professional condition-monitoring services. Such services already have a staff of trained technicians who are engaged in predictive activities every day. Have them conduct an on-site presentation of their machine-monitoring services. If economically feasible, it's a place to start.

The question has been asked many times, and it's worth repeating: "Where do you start implementing PM?" Focus first on critical production machinery and equipment, plus large motors, generators, compressed air, process supply systems, and other powerhouse applications. After that, remember the 80/20 rule, because it probably applies here: 20 percent of the machines cause 80 percent of the problems. Look at the 20 percent, using the machinery and equipment owner's manuals as the primary resource for determining what PM tasks to perform.

Be conservative. Don't add too much PM too quickly. Make sure the process is working correctly first before expanding PM activities.

PM should be integrated into production scheduling. All PM that can be performed during operation of the machines should be done at that time. It's important where possible for production to actually see the PM work being performed. It's management's role to ensure that a plant's PM work is being routinely scheduled, correctly performed, and completed on time.

Observations about Preventive Maintenance

PM may be one of the most misunderstood and misapplied of all of the tools of maintenance management. Too often, it's too much or too little or too late or not at all! All physical plant assets require routine maintenance service as defined in the owner's manual provided by the original equipment manufacturer (OEM). Remember these truths about PM:

- If it doesn't prevent something, it isn't PM.

- PM is not an acronym for *Perform Miracles*.

- The success of PM is dependent upon measurable results.

- PM *must* economically justify itself, or it will become a victim of austerity.

- The most cost-effective form of PM is to procure machinery and equipment that have the greatest reliability and maintainability. This reduces the need for maintenance.

- Proper lubrication is one of the most cost-effective forms of PM. The major reason it is so cost-effective is that it's so terribly expensive if neglected.

- Focus PM first on critical machinery and equipment. That means machines operating at capacity and high-usage machines that have no backup.

- After the critical machines, use the 80/20 principle to identify the 20 percent that cause 80 percent of the problems and PM them.

- One of the first efforts in PM should focus on the proper application and operation of production machinery and equipment.

- Be conservative; too many PM programs have failed due to overkill.

- The measurement of success for PM is the cost of machine malfunction and failure.

- PM truly is a matter of $ and Sense.

Chapter 8
Maintenance Spare-Parts Control

Spare parts, materials, and supplies are an important element of maintenance cost; it's approximately equal to the cost of maintenance labor. Unfortunately, serious problems exist in the management of these maintenance materials. Over the years, spare-parts-related problems have been a major contributor to excessive maintenance costs, machine downtime, and related production costs and losses. In spite of this, the same kinds of problems have continued unresolved for decades. Issues include the following:

- major investment in idle inventory
- costs of maintaining idle inventory
- uncontrolled inventory in the plant
- ordering materials already in-house
- continuing outages and shortages
- duplicate setups and orders
- expediting and premium freight
- waiting and other delays at the crib
- excessive travel to and from the crib
- lost time identifying parts/part numbers
- lost productivity of maintenance workers
- associated machine downtime/costs and losses
- negative impact on bottom-line profits

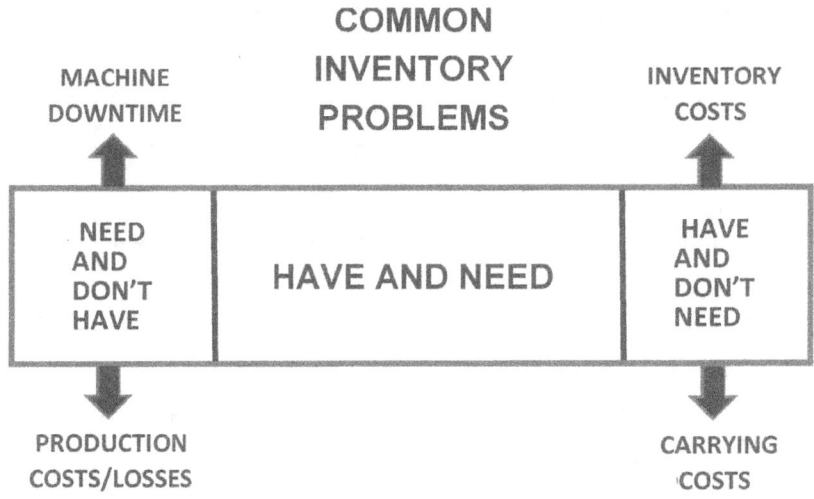

Many (if not most) maintenance managers, foremen, and skilled-trade employees consider the continuing outages and shortages of spare parts and other materials to be their biggest problem. However, the systemic problems involved are more complex than that; they are rooted in dysfunctional structure, systems, processes, policies, procedures, and practices. This theme of dysfunctional management controls has been repeated over and over, because this is where the majority of problems are found in large social systems such as a manufacturing plant; it's also true in the managing of maintenance spare parts.

As with other parts of maintenance management, the best way to understand the root causes of problems and potential solutions is to begin with the basics. It's important to know that maintenance spare parts, materials, and supplies are classified as maintenance repair and operations (MRO) materials. MRO materials are tax-exempt expense items intended for short-term use. In reality, *short-term* is a misnomer, because while some MRO materials are used immediately upon receipt, the majority of these items must be held in inventory ready for immediate use to avoid costly extended machine downtime.

Managing the in-house inventory is a maintenance-stores function. The place where MRO materials are stored is commonly called the maintenance *crib*. In most plants, floor space comes at a premium, so the maintenance crib often lacks adequate space to properly store all of its inventory. As a result, these MRO materials can literally be found all over the plant—in basements, under stairwells, in mezzanines, in boiler rooms, in out-buildings. Anywhere there is a nook or a cranny, MRO materials can be found. Rarely is there any documentation of the location of these items.

The consequences of this out-of-control situation are predictable and include:

- ordering materials that are already in the plant ... somewhere
- avoidable machine downtime waiting on ordered parts that are already in the plant ... somewhere
- new MRO materials stored out in the plant becoming unusable due to corrosion, moisture, or other effects of being improperly stored

The monetary losses involved can be huge, and the costs of correcting this situation can be huge as well, in terms of the time and resources needed to create a secure space for storing the materials; finding and moving the materials to the secure location; identifying the materials by manufacturer, part number, etc.; and cataloging the items. Depending upon the magnitude of the problem, even with dedicated resources, this may take several months.

The following are items typically stored in the maintenance crib:

- spare parts
- materials
- supplies
- tools
- equipment
- miscellaneous

Historically, the role of crib attendants has included the following:

- setting up
- cataloging
- coding
- ordering
- receiving
- stocking
- dispensing
- accounting

Three different manual methods of inventory control have been used in the past. The first is called *one for one*. Using this method, each time an item is dispensed, the same item is ordered. The second method is called *dual-bin*. With this method, inventory is equally divided into two bins. When the first bin becomes empty, that bin quantity is ordered, and the second bin is put into use. The objective is to refill the first bin before the second bin is emptied. This process keeps repeating itself. A third method is called *bin count*. Using this method, the crib attendant periodically counts the items remaining in the bin. At a predetermined level, the required quantity is ordered to refill the bin. All three methods have been used depending on the type of items involved.

Computerized inventory control adds two additional options. The first method automatically creates an order based on *reorder point* (RP) and *reorder quantity* (RQ). RP equals a specified minimum-on-hand quantity that automatically initiates the order. RQ is the order quantity required to achieve the maximum-on-hand level. The second method employs computer (logic) algorithms to determine RPs and RQs. A simple example of this might involve ice-melting salt. The following is an algorithm presented in text form rather than in a computer language or mathematical formula:

> (If the Month is November, December, January or February use the established 'Reorder-Point' and 'Reorder-Quantity'. If the Month is March, April, May, June, July, August, September or October, do-not-order.)

Obviously, situations can arise that require on-the-fly adjustments to these otherwise automated systems. The fly in the ointment that can't be controlled by computer is the vendor delivery element. That's an entirely different issue that also must be managed.

In some plants, the inventory control of MRO materials could be performed using the production materials management system. Where that isn't possible, there are commercially available systems designed for use by the maintenance department. As with any operational function, there are objectives. In the case of MRO materials, there are two key objectives, and they must be properly balanced.

Cost Objectives	**User Objectives**
Minimize investment	The right material
Maximize inventory turns	The right quantity
Maximize service levels	The right location
Minimize staffing	On time every time

While both sets of objectives are legitimate, they can be in conflict with one another. Management must be especially careful that the cost objectives of the maintenance stores operation does not result in extending machine downtime (and it can), with far greater production costs and losses.

It's during times of austerity that conflicts between costs and MRO material availability is most likely to occur. To those who live in the financial world of costs, budgets, and forecasts, MRO materials may look like a good place to reduce costs by delaying spending, reducing order quantities, etc. Unfortunately, you cannot arbitrarily reduce machine failures that require the use of MRO materials. The maintenance manager should become proactive and initiate a comprehensive review of MRO inventory types that could safely be reduced or delayed without compromising production machine uptime objectives.

Preparing this list ahead of time may be the best way to avoid an across-the-board reduction in MRO material spending. Also, it would be a good idea to review the plan with production management; their support would be a plus.

In-plant surveys have shown that maintenance employees spend a significant amount of nonproductive time in such MRO materials-related activities as the following:

- searching for materials
- trying to identify materials
- substituting materials when the needed item isn't available
- traveling to and from the jobsite to obtain materials
- waiting at the maintenance crib to obtain materials
- transporting materials

It has been estimated that these nonproductive activities may consume an hour or more of each employee's eight-hour workday. It's important to focus on the maintenance crib and its operation because it is inexorably linked to the effectiveness, efficiency, and utilization of a plant's skilled maintenance workforce. There's a need for a new maintenance-crib paradigm.

MRO parts and materials are identified with part numbers. The manufacturer of the part provides a part number for ordering purposes. When the part is put into inventory in the plant, it may get a new part number for inventory control purposes. Crib attendants use part numbers; they think part numbers. People who work with part numbers all day can remember part numbers, or logical coding schemes, or proper noun names, but don't expect the repairman to remember. For example, is the correct name *actuator, bar, connector, handle, lever, lifter,* or *rod*? It's not as simple as one might think.

Since inventory control requires the use of part numbers, the crib catalog becomes the bridge between part names and part numbers. This would seem to solve the problem, except there have been too many examples of catalogs being designed for the crib attendant rather than the casual user: the maintenance worker. The catalog needs to be designed with time in mind. How quickly can the casual user identify the part number? The repairman who is working on a machine that is down needs that information as quickly as possible in order to access the needed part. That requires the catalog being structured so that it is intuitive to the user. The only way that's going to happen is by involving users in designing the catalog.

Obviously, there is no one way to structure a catalog. However, there are examples that can be used as thought-starters. Since computerized catalogs have become so common, the following examples will be presented as computer screens. Each computer screen display is like a page in a paper catalog.

The screen below represents the home screen for MRO materials:

```
┌─────────────────────────────────────┐
│          MRO MATERIALS              │
│            ELECTRICAL               │
│        BEARINGS & SUPPLIES          │
│            HARDWARE                 │
│          HOSE & FITTINGS            │
│          PIPE & FITTINGS            │
│        POWER TRANSMISSION           │
│         REPLACEMENT PARTS           │
│        STRUCTURAL MATERIALS         │
│          TOOLS & EQUIPMENT          │
│           MISCELLANEOUS             │
└─────────────────────────────────────┘
```

Assuming that the display is a touch screen, a selection can be made simply by touching the word desired. This demonstrates the use of names instead of numbers. If a 90-degree elbow is needed, touch *PIPE & FITTINGS*, and the following screen is displayed:

PIPE AND FITTINGS

ADAPTERS	LOCKNUTS
BUSHINGS	NIPPLES
CAPS	PLUGS
CLAMPS	PIPE
COUPLINGS	REDUCERS
CROSSES	TEES
ELBOWS 45	TRAPS
ELBOWS 90	UNIONS
FLANGES	VALVES
HANGERS	Y-LATERALS

Touch *Elbows-90*, and the following screen will be displayed:

SIZE in INCHES

1/8	1-1/2
¼	2
3/8	2-1/2
1/2	3
3/4	4
1	6
1-1/4	8

Touch *1inch* to display the following screen:

```
      PART NO.          ELBOW 90 1-INCH
         62323 Forged Steel - 3000 Lb Flanged
         20027 Galvanized Steel threaded
         20048 Galvanized Steel Threaded Street
         347242 Malleable Iron Schedule 80 threaded
        1004326 PVC Schedule 80 Socket
        1004328 PVC Schedule 80 Threaded
        1004492 PVC Schedule 80 Street-ell Threaded
        1004496 PVC Schedule 80 Union-Type Threaded
        1005033 CPVC Schedule 80 Threaded
         297432 SS Type 304 150 Lb Threaded
         352753 SS Type 316 150 Lb Threaded
```

The above screen lists all of the 1-inch 90-degree elbows in inventory. This allows the requestor to select the item desired or a possible alternative if the item wanted is not in stock. If *1004326* is selected, the following screen is displayed:

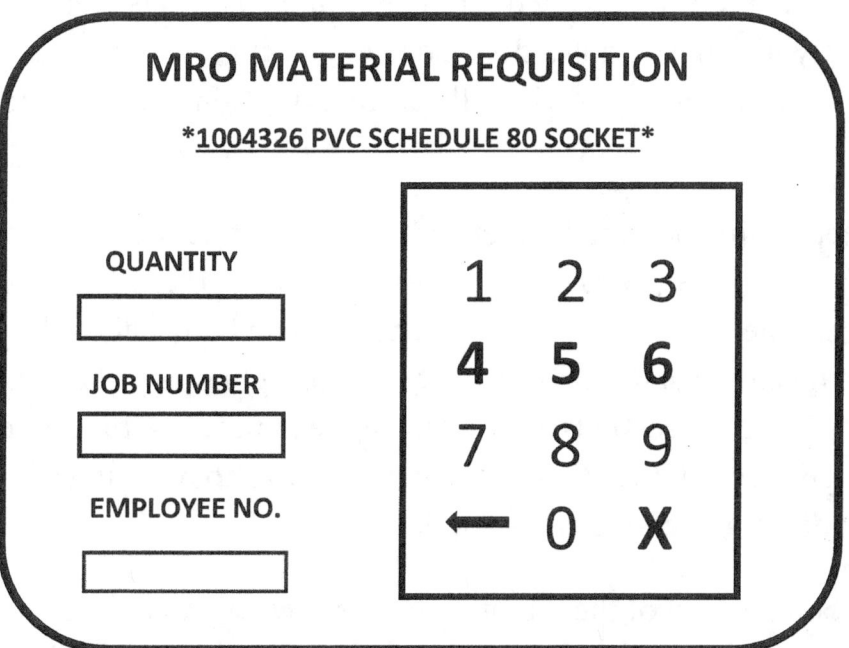

This screen creates an electronic requisition that documents the quantity to be withdrawn, the job number, and the employee number.

The job number has (in the background) the machine number where the part is being used, the department where the machine is located, the cost of the part or parts, and the account to be charged.

This type of catalog provides for detailed management reporting of MRO material usage by part number, machine, and department. It also continuously updates usage for inventory-control purposes. The time required for maintenance employees to learn to use an intuitive electronic catalog such as this is minimal and would significantly reduce the time required to access and requisition part numbers. This is especially important when saving time reduces machine downtime and the associated production costs and losses. Such a system in a large manufacturing company would have a very attractive return on a one-time investment in skilled-labor savings that would continue on, year after year.

While it's important to effectively control inventory and use technology to speed up the process in ways that improve worker efficiency, there are other aspects of MRO materials management that need to be considered. It should be remembered that modernizing the maintenance crib operation is an important step in the larger effort to minimize the total cost of maintenance. The next obvious step in this process is to incorporate the use of bar coding. This has major cost benefits: it significantly improves transaction speed and accuracy, reducing wait time for the maintenance customer and virtually eliminating common manual transaction errors.

The first step in this process is to ensure that all vendors provide bar codes for the items they ship. It's probable that most already do. If the vendor's bar code is different from the crib's, a crib bar-coded label can be printed when the item is scanned for receiving at the crib. Using bar code technology changes transaction times from minutes to seconds. Considering the number of transactions that take place during each shift, over time that can translate into reducing a lot of machine downtime and its associated costs and losses.

There is another aspect of the maintenance crib where a paradigm shift is badly needed: the physical structure of the crib itself. The current physical structure is appropriately called the grocery-store concept. The typical grocery store in the US from the 1800s through the 1950s commonly had a counter at the front of the store that separated the customers from the groceries. The customer would either provide a list or tell the grocer what was needed and then wait at the counter while the grocer filled the order. If there were other customers, they waited in line for their one-at-a-time turn. It didn't take long for frequent customers to learn where most everything in the store was located. They could have easily and quickly gotten what they needed themselves except for the sturdy counter.

However, something happened in the 1940s and 1950s to change all that: supermarkets came on the scene. When customers entered the store, they had free access to all of the store's stock. They filled their own order and paid for it at checkout. A new paradigm was born.

The same paradigm can be applied to the maintenance crib: free (controlled) access for maintenance employees, allowing them to get their own parts and materials so that multiple people can get what they need at the same time. The exit would be controlled by the crib attendant so everything removed from stock could be properly documented. The use of bar codes creates a much faster in-and-out process.

The next page shows an example of the grocery-store concept that has been in use for so long. The following page show an example of how that same crib might look using the supermarket concept. These are not suggestions, only examples to demonstrate the changing paradigm.

THE CURRENT PARADIGM
(The 1850-1950 'Grocery Store' Concept)

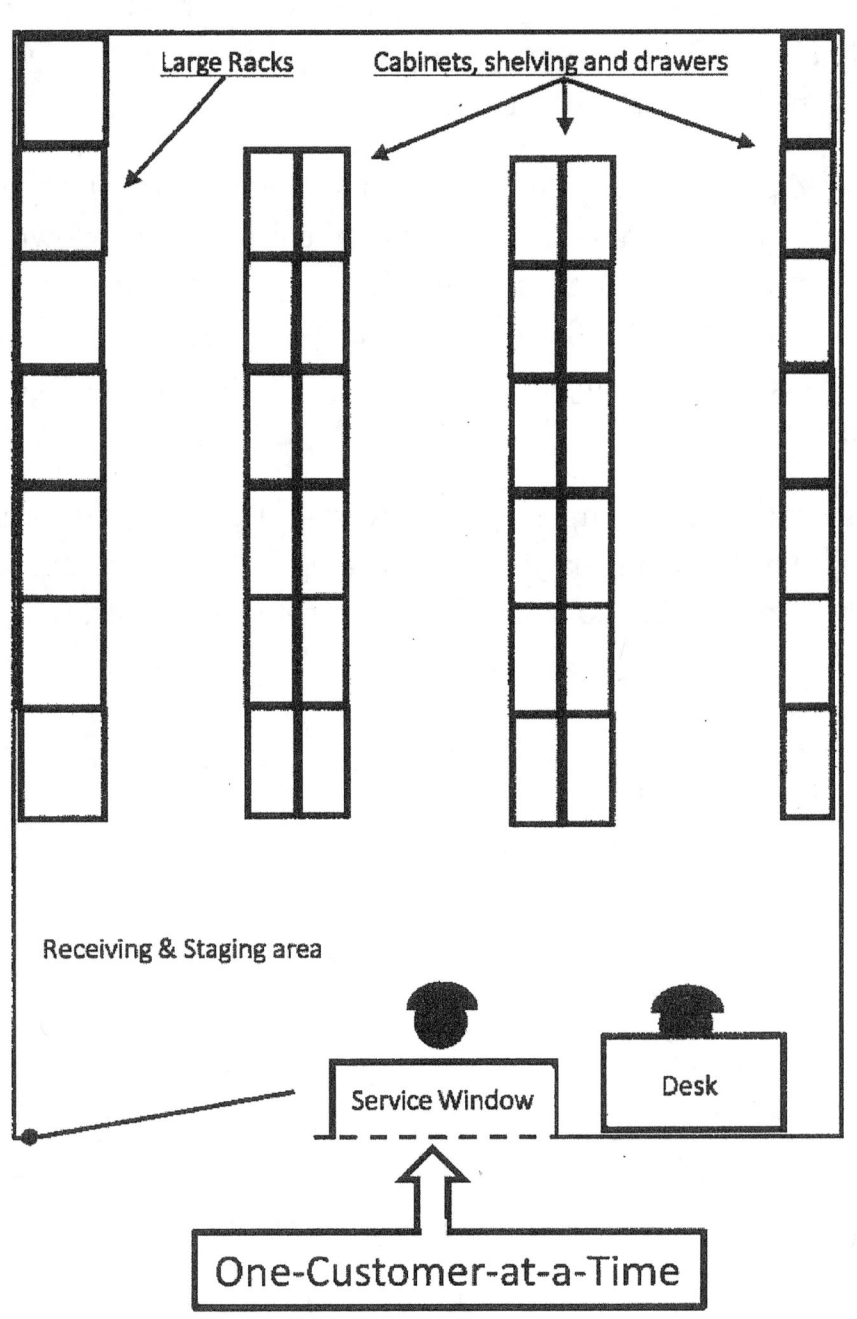

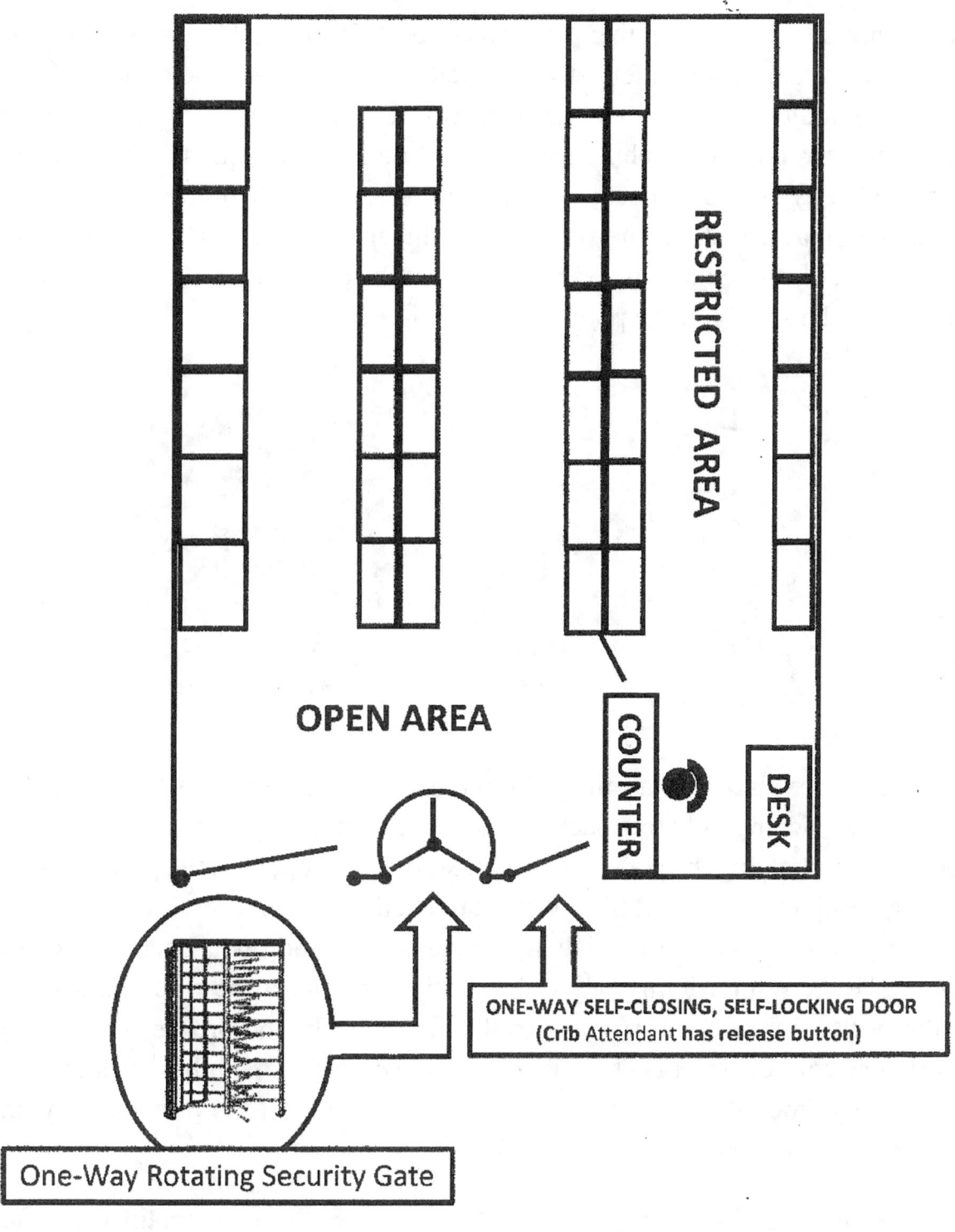

Note that the supermarket crib has a restricted area where material requiring special authorization is stored. This might include things like tools or other high-value materials. It might include electronics that require special handling and storage.

There are other aspects of the new crib concept. First of all, there would be the central crib for keeping safety stock as well as large, heavy, and seldom-needed items. In addition, there would be point-of-use mini-cribs strategically located in specific production machine areas. These might be used for storing parts that are needed by only one machine, or placed in an area where there are multiple machines that use the same parts or a critical machine where there is high downtime. Shown below are examples of the type of lockable storage equipment that might be suitable for a point-of-use mini-crib.

For each part placed in a mini-crib, there is an accompanying pre-filled-out requisition lacking only the machine number and quantity, to be filled in by the person withdrawing the part. These requisitions are kept with the part inside the storage cabinet. There will be a pocket attached to the outside of the cabinet for placing requisitions for materials that have been withdrawn. Each workday, it would be the responsibility of a crib associate to collect the filled requisitions from each mini-crib, return them to the central crib, and process them in the inventory control system. The crib associate would then withdraw from central crib safety stock the same part numbers that were used the previous day and restock the mini-cribs.

In order for this system to work, the tradesmen who have complained so often about parts being out of stock will need training to understand how the system works, emphasizing their responsibility for writing in the machine number and quantity withdrawn on a requisition and then placing the completed requisition in the proper location. Otherwise, no parts will be ordered to replace the withdrawn part. When people understand the benefits, they *will* make the system work.

The use of crib personnel to perform nontraditional duties outside the crib is an important part of the new paradigm, which transforms plant maintenance from what it has been in the past into a far more effective and efficient operation that contributes to the objectives and goals of the people it serves. In order to accomplish this, the traditional role of the maintenance crib attendant will change, and this may require a name change—MRO materials associate, for example. If this requires a change in pay scale, that can be justified based on increasing the effectiveness and efficiency of the skilled-trade workforce. The objective is to make the former crib attendants production support team members. They need to feel the same sense of responsibility and urgency to restore a failed production machine as the skilled-trade people do.

One nontraditional way in which this could be accomplished is to include the maintenance crib personnel in the use of two-way radios, as explained in the chapter on maintenance dispatching. This would enable an electrician working on a high-priority down machine who has identified a failed switch as the problem to call the crib via two-way radio and have the part delivered to the jobsite while the failed switch is being removed, thus eliminating the need for the electrician to stop work on the down machine, travel to the maintenance crib, wait to obtain the new switch, then return to the jobsite. Obviously, machine downtime would be reduced in this instance; but think about how much downtime could be reduced over time using this concept.

Remember, the new paradigm for spare parts control is the result of making appropriate changes to organizational structure, systems, processes, policies, procedures, and practices.

Chapter 9
Priorities
The Achilles Heel of Maintenance and Production

All of us have priorities that govern our lives. People make decisions every day based on their priorities—perhaps doing so without even thinking about priorities. Priorities are just a part of life's experience.

Priorities are also a fact of life in a manufacturing plant, beginning with the plant manager and extending to every supporting entity in the plant, including production and plant maintenance. The production department and maintenance department are linked by the priorities established by production. However, when production requests maintenance assistance, it must provide enough information so that maintenance will send the right resources.

Maintenance will also need to know how important the job is. What's the priority? That's because there are periods of peak demand when there are multiple machines down and all of the maintenance resources are currently working on other jobs. The department will need to know how important the new job is in relation to the jobs already in progress. It's important to point out that it's not possible for the maintenance department to know the current (and often changing) priorities of production. That's the job of the production department.

Unfortunately, there are far too many plants that don't have any kind of formal structured priority system for the maintenance of production machines. Even when there is some type of priority system, priorities often get abused. When that happens, lower priorities can get preference over higher priorities, and that can become costly to production.

Who would abuse priorities? The production foreman, that's who! The foreman's performance is being measured by how well certain production objectives are met, and a foreman will do whatever it takes to meet those objectives. When machines stop, production stops. Getting that machine running again is the foreman's highest priority, even though it may not be the plant's highest priority.

Don't blame the foreman for trying to get the job done; but there must be controls for a priority system to be successful. There is a solution: it's called a condition-based priority system. This has been used successfully in other manufacturing plants, and that success can be replicated. Unfortunately, it isn't a one-size-fits-all off-the-shelf ready-to-be-installed product; it's more of a do-it-yourself project, and that's what makes it successful.

The following structured condition-based priority system for plant maintenance was developed in-house at an electronics manufacturing plant using a professional facilitator. This plant was also operating a centralized maintenance dispatching system. The first thing in the process was to define what would be the highest priorities of all: emergency situations. Examples include the following:

- employee or employees trapped or injured
- fire
- explosion
- structural collapse
- ruptured gas or chemical lines
- electrical shock
- hazardous material spills
- severe weather emergency

Since these situations are of the highest priority, they need to be identified uniquely so that they cannot be confused with other priorities. Hospitals use a color code to identify emergencies. Code red was a good choice for this manufacturing plant.

Normally, these situations are already a part of a plant's emergency response plan. If a code red is communicated to the maintenance department, it immediately dispatches a emergency response team to the scene. There is also a class-1 priority, which is the highest priority for response to machine malfunctions and failures. This priority is reserved for what has been identified as a critical machine. A *critical* machine is one that has been classified critical by the plant manager or production manager. Critical machines include:

- machines operating at or near rated capacity
- machines such as heat-treat with no backup

- fully scheduled specialized machines with no backup
- test equipment with no backup
- centralized process supply systems
- plant safety and security systems

A class-1 priority is established for any critical machine that malfunctions, resulting in any of the following conditions:

- not functioning
- not producing quality parts
- not operating at scheduled output rate
- operating with obvious problems
    - unusual noises
    - unusual vibration
    - excessive heat
    - smoke

The maintenance department is committed to a maximum five-minute response time for getting repair people to the jobsite.

A class-2 priority is established for all noncritical machines that malfunction, resulting in any of the following conditions:

- not functioning
- not producing quality parts
- not operating at scheduled output rate
- operating with obvious problems
    - unusual noises
    - unusual vibration
    - excessive heat
    - smoke

The maintenance department is committed to a maximum fifteen-minute response time for assigning and getting someone to the jobsite.

A class-3 priority is established for all noncritical machines that are currently operating but malfunctioning due to the following conditions:

- reduced scheduled output rate
- creating minor scrap or rework
- creating direct-labor inefficiency

Maintenance will respond to class-3 priorities as soon as all higher priorities are satisfied.

In order to meet response-time commitments, the following logical order is established for making job assignments.

Assignment Logic for Class-1 Jobs

- Assign to available employee.
- Reassign employee from class-3 job
- Reassign employee from class-2 job
- Assign backup employee.

Assignment Logic for Class-2 Jobs

- Assign to available employee.
- Reassign employee from class-3 job.
- Assign backup employee.

Backup employees are those currently assigned to facility maintenance or scheduled routine maintenance.

The control of priorities is one of the most cost-effective of all management activities. It's essential that machine maintenance priorities be institutionalized into the production system. The lack of effective priorities is one of the greatest single contributors to (avoidable) production costs and losses in manufacturing plants today. Therefore, people need to be made accountable for proper priority decisions.

There are two things that can be done to ensure effective placement of priorities. First, make people accountable for results by adding the proper use of priorities to the employee appraisal process. Second, use the machine maintenance reporting system to list priorities by department, machine, and date. This would be submitted to the production manager.

The information provided here should not to be considered *the* solution; rather, it's an example, a thought-starter for creating an in-plant-developed solution. If the users of the system help develop the system, the probability of success is enhanced by orders of magnitude.

Remember, priority control is cost control.

Chapter 10
Organizational Structure

Organizational structure is the supporting framework that gives the organization form and function. There are two major types of organizational structure: management structure and operational structure. Management structure defines key attributes of the organization as superior-subordinate relationships, peer relationships, and levels of authority.

Management structure is displayed visibly on the company's organization chart. There is no single best way to structure management for plant maintenance. The proverbial maintenance manager who has responsibility for the plant's maintenance operations has been shown at many different levels on company organization charts. There have been instances where the maintenance manager has functioned in the role of a first-level foreman, supervising hourly employees. There have also been cases at the other end of the spectrum where maintenance managers are in a staff position, having multiple levels of management reporting to them. The differences are commonly linked to the size of the company or its parent corporation.

Regardless of the position, title, or where they are listed on the company organization chart, maintenance managers are where the buck-stops-here responsibility lies for plant maintenance operations. Senior management, with input from maintenance professionals, is probably in the best position to determine the management structure of plant maintenance operation, although industry precedents may offer some useful perspectives.

The operational structure of plant maintenance takes many forms that have made a major impact on the effectiveness and efficiency of maintenance services. How maintenance structures itself to provide service to its production customers will strongly influence just how well or how poorly maintenance activities meet customers' objectives. After all, isn't that the purpose of plant maintenance—to meet the needs and goals of its customers?

Employee Travel-Time Issues

Before proceeding with the structuring of plant maintenance operations, there's an important factor that should be considered and probably never is: where in the layout of the plant has management located the maintenance shop and parts crib? The configuration of the building, the placement of production machinery and equipment within the building, and the location of the maintenance department automatically determines the distance maintenance employees must travel to and from a jobsite. Considering how many jobs will be performed by the maintenance workforce over the course of a year, every year, that may be a lot of traveling.

The significance of travel time cannot be overemphasized. Walking isn't tool-in-hand work that's getting a failed machine up and running again. In fact, it's just the opposite. It's delaying work; it's adding to machine downtime.

The following exercise uses a drawing to simulate a plant layout, plus some mathematics, to help to clarify the significance of travel time and demonstrate why the location of maintenance facilities can matter a lot. The drawing below (which is not to scale) represents the floor plan of a 1-million-square-foot production facility. This building is 800 feet wide east to west and 1,250 feet long north to south.

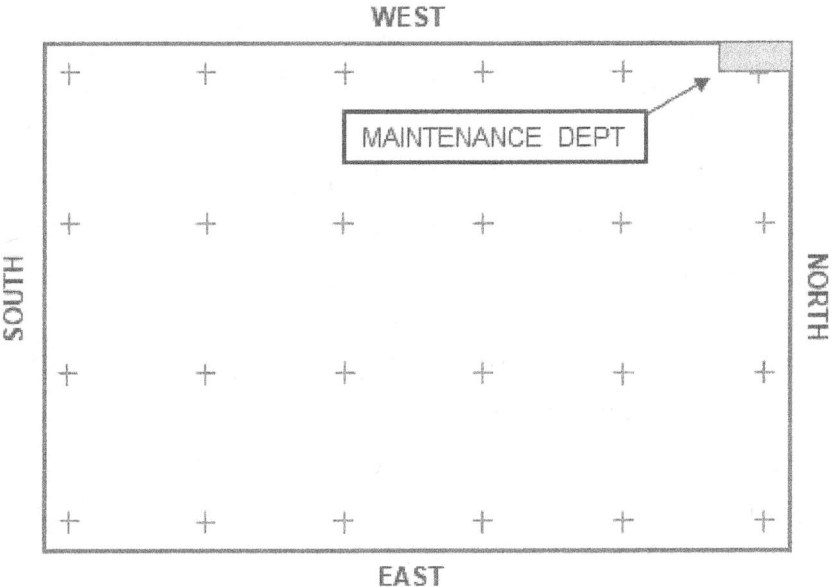

To facilitate this exercise, we'll say that the plant has twenty-four maintenance employees working one shift, and there are twenty-four jobsites distributed somewhat equally throughout the building. In addition, the maintenance department's operational structure is centralized. All requests for maintenance service are routed to the maintenance department and assigned out of that facility. This data was established to make it easy to calculate travel times and distances; it has no influence whatsoever on the validity of the example.

The twenty-four jobsite locations inside the building have been marked with a + symbol. These jobsites are configured six rows east–west and four rows north–south. The perimeter jobsites are all located fifty feet from the closest wall. Since the building is 800 feet wide east–west, the jobsites located along the east wall are 700 feet from the jobsites located along the west wall. Similarly, since the building is 1,250 feet in length north–south, the jobsites located along the north wall are 1,150 feet from the jobsites located along the south wall. The six east–west rows of jobsites are equally spaced 233.3 feet apart. The four north–south rows are equally spaced 230 feet apart. This configuration establishes a precise, measurable location for each jobsite within the building, making it possible to calculate the distance from the maintenance department to any of the twenty-four jobsites.

The building is filled with a large number of machines and other equipment. Traditionally, these assets are installed parallel or perpendicular to the building's walls. After the machinery and equipment are installed, designated aisles will be painted on the plant floor; they too will be parallel or perpendicular to allow safe travel between machinery and equipment. Maintenance employees can't travel through the plant as the crow flies because of the machinery and equipment; they must use the designated east–west or north–south aisles to reach the jobsites. Since the maintenance department is located at the northwest corner of the building, with the exception of the jobsite closest to the maintenance department, all travel from the maintenance department to jobsites will be east or south or both.

For example, if John is dispatched from the maintenance department to the jobsite at the far southeast corner of the building, no matter what route he travels to get there, he will have to walk 700 feet east and 1,150 feet south to reach the jobsite. The total east and south travel distance is 1,850 feet. Upon returning to the maintenance department, he will have walked a total of 3,700 feet. On the other hand, if John were to be dispatched to the jobsite next to the maintenance department, the walking distance would be zero.

For the purposes of this example, if all twenty-four employees are assigned, one to each of the twenty-four jobsites, the total two-way distance traveled by the twenty-four employees would be 22,199 feet, or 4.2 miles. If the maintenance department was located at the north–south east–west center of the building, the same twenty-four people traveling to the twenty-four jobsites will travel only 13,878 feet, a reduction of 8,321 feet, or 37.5 percent less travel.

In addition to distance, there's another important dimension to travel: time. The reason travel time is important, in many if not most cases, is that it extends machine downtime, and machine downtime translates into associated production costs and losses. The longer the machine is down, the greater the costs involved.

A recent Google search provided the following information from the *Wikipedia* page about walking:

> The average human walking speeds at crosswalks is about 5.0 kilometers per hour (km/h) or about 1.4 meters per second (m/s), or about 3.1 miles per hour (mph). The average time it takes most people to walk a mile is about twenty minutes (walking at about 3 mph).

Using this information to provide a worst-case scenario about how walking contributes to machine downtime, let's revisit John's trip to the southeast corner of the plant, starting at the point where he arrives at the jobsite. He has already walked 1,850 feet. The machine is down and production has stopped. It takes John about thirty minutes to find the problem: a broken gear. He returns to the shop for a new part. At this point, he has walked 3,700 feet.

With the new part and the necessary tools in hand, he returns to the jobsite. The distance traveled is now 5,550 feet. When the damaged gear is removed, John discovers that the broken gear has damaged the gear behind it; this couldn't be seen until the first gear was removed. There is nothing to do but make a second trip to the maintenance department. The total travel distance is now 9,250 feet.

It takes about an hour and a half to replace both gears. When everything is ready to go, the machine operator starts the machine. Everything seems okay, but there's a popping noise. The operator states that he has heard the noise before but didn't think too much about it. With the machine turned off, John soon finds the problem: a timing belt is beginning to come apart and is flapping against a guard. The production foreman is called to look at the problem, and John warns him that the belt could break at any time. The foreman agrees and tells John to replace it.

The two-way trip to the maintenance department for a new belt adds another 3,700 feet of travel. The total is now 12,950 feet, and John is getting tired. Regardless, the new belt is quickly installed, and the machine is restored to production.

Using the Wikipedia data of 20 minutes per mile, travel time alone has extended machine downtime by 49 minutes. How much did those 49 extra minutes cost production, and how many 49-minute situations like this will occur over time? This example was presented as a worst-case scenario, which may not happen very often, but it does actually happen periodically.

There are things that can be done to help mitigate these situations. For example, when John had to make those long walks back and forth to the farthest corner of the building, wouldn't a three-wheeled bicycle with a rear-mounted toolbox have made a wonderful difference, both to John and the machine downtime involved?

This worst-case scenario was only one job in a plant where there will be a multitude of jobs. Another example can better demonstrate what actually takes place in large discrete manufacturing plant. Using data from the worst-case scenario, a far more realistic scenario is created to demonstrate the significance of travel time.

It was previously calculated that having all 24 maintenance employees assigned, one to each of the 24 jobsites, would result in 22,199 feet of travel. The average travel distance for the group would be 925 feet of travel for each employee. Based on experience, a conservative estimate for a large manufacturing plant would be three or more jobs per worker per day. Using an estimate of 3 × 925 feet, each employee will travel an average 2,775 feet per day. Combined travel for all 24 employees is 66,600 feet per day. Based on a standard work year of 240 (no overtime) days, the annual travel for the 24 employees is 15,984,000 feet, or 3,027 miles per year.

Using the Wikipedia data (3 miles per hour), 3,027 miles will require 1,009 hours of walking each year. At $25/hour, that's $25,225 every year for nonproductive walking. What's worse, at least 50 percent (500 hours) of that walking will contribute to 500 hours of machine downtime. What will 500 hours of downtime add to the cost of production each and every year unless something is done to mitigate the situation?

Some may totally discount what has been presented. Others may feel that the estimates are too high. That's possible, but they also may be too low. The purpose of this exercise is to focus on the issue of excessive nonproductive walking by a plant's skilled maintenance workforce. However, the issue should not be viewed as a big problem. Rather, it should be viewed as a big opportunity to bring about positive change, because there are so many things that can done to reduce nonproductive walking. Positive change can be brought about through changes in operational structure.

Structural Focus

There are a variety of different structures that not only reduce excessive walking but also improve the utilization, effectiveness, and efficiency of skilled maintenance resources. Each of these structures is presented with its advantages and disadvantages.

Skill-Focused Structure

This is an outgrowth of the establishment of union-certified craft skills, such as electrician; carpenter; welder; tool and die maker; sheet metal worker; and pipefitter. For example, an industrial electrician has a highly technical skill, so a foreman over electricians must have electrical skills to properly supervise electricians, right? That may or may not be true.

There certainly are some advantages to having an electrician supervising electricians. Such a foreman can provide technical assistance to the employees and assess individual training needs, plus absence has less of an impact on a group of the same skill set. There also are disadvantages, in that rarely if ever will the foreman have been trained in interpersonal skills or even the most basic business skills. More often than not, it's a matter of learn as you go if you can. Additionally, many jobs will have multicraft requirements. It becomes more difficult and time-consuming to coordinate multicraft jobs, and who has primary responsible for them? Also, if the supervisor is skilled, there may be a tendency to micromanage by supervising the work instead of the people.

Is the company better off having a foreman with good technical skills or good supervisory skills? Obviously, both would be preferable, but how many such foremen are out there waiting in the wings? It has been demonstrated again and again over time that supervising a group of technical people does not require the same technical knowledge as those being supervised if the supervisor has good interpersonal skills, and especially if those are complemented by good business skills.

Job-Focused Structure

This structure is focused on the foreman having a mixed group of people with different skills. It has several advantages. When coordinating multicraft jobs, the foreman has all the skills needed to get the job done. All of the employees work for the same foreman, and he's responsible for the job. It's much easier for such a foreman to put together a multicraft job crew. It facilitates a team concept and a greater degree of mutual helpfulness.

This is the right way to structure when using the area maintenance concept. The one big disadvantage is the negative impact of absence.

Other Structures

There are other structural types that can have significant impact on the effectiveness, efficiency, and utilization of the maintenance employees who support the production areas of the plant. Several factors must be considered when determining the best way to provide maintenance service to production, including the following:

- physical size and layout of the plant
- quantity, diversity and complexity of machines
- size and makeup of the skilled workforce
- work rules and lines of demarcation
- communication capabilities

Operational Structure

There are four major types of operational structure currently seen in manufacturing plants: centralized, area, flexible area, and production teams. We'll look at each in detail.

Centralized Maintenance

The centralized maintenance structure has probably been around since manufacturing plants first appeared on the scene. Using this concept, requesters notify the maintenance department and employees are dispatched from there to the jobsite.

I am reminded of a scene out of the twenties or thirties where the plant manager takes his daily tour of the plant. His first stop is to look in on the maintenance department, where he sees people playing cards, reading, or even sleeping; he turns away smiling, feeling confident that everything is up and running smoothly out in the plant.

While that may seem odd to some, it wasn't uncommon at the time because the concept of preventive maintenance (as we know it today) had yet to appear. Nevertheless, using the centralized maintenance structure (in the absence of wireless communication), maintenance workers still routinely come back to the maintenance shop for a new job assignment. Unfortunately, in some cases, that means they then have to return to the vicinity of the job they just completed.

'Centralized' Maintenance

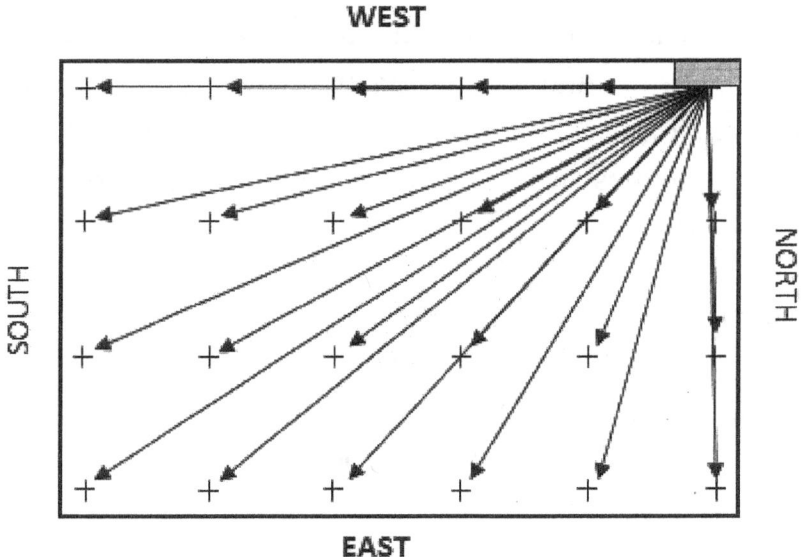

Advantages of Centralized Maintenance

- maximum flexibility in the use of the workforce
- most effective use of craft foreman
- easier to manage plant-wide priorities
- maximize use of shop facilities
- develops a wider range of work experience
- least impact of employee absence
- easier to administer balance of OT hours

Disadvantages of Centralized Maintenance

- longer response time due to travel time
- less in-depth experience with specific machines
- more difficult to coordinate multicraft jobs
- accountability for multicraft jobs becomes vague
- more lines-of-demarcation problems
- less sense of partnership with customers
- less accountability for specific machines/processes

Area Maintenance

This operational structure was probably the first departure from the traditional centralized maintenance approach. Using this type of structure, the plant is divided into a number of areas (sometimes called zones) where a maintenance foreman and a multicraft group of employees become responsible for the machinery and equipment within that specific area of the plant. Sometimes the configuration (size and shape) of the plant becomes a factor in creating areas. At other times, the area might be configured to accommodate product lines or processes. The crew size and craft mix for these areas are (hopefully) matched to the types of machinery and equipment in the area.

'Area' Maintenance

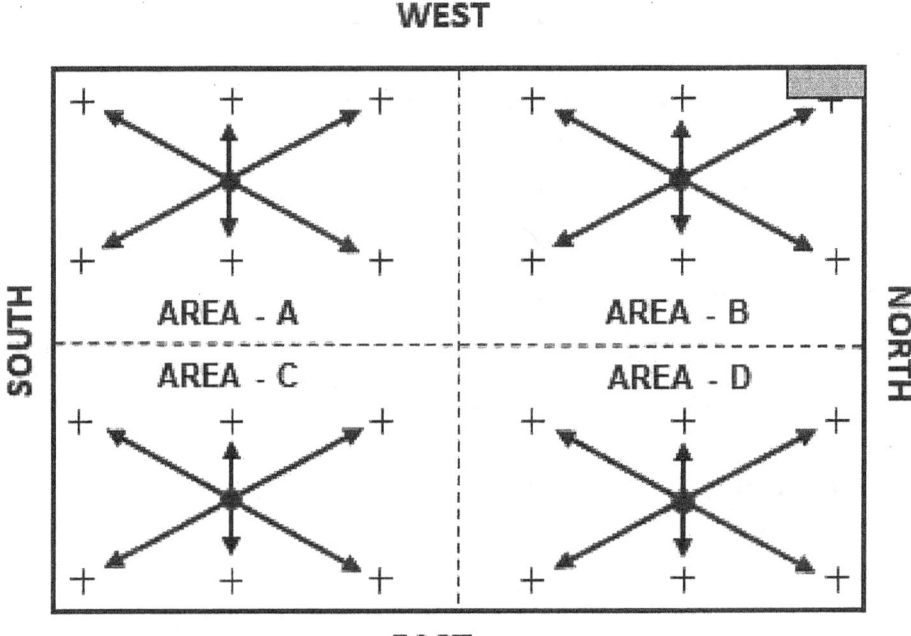

The first-glance impression, looking at the layout, is of dramatically reduced travel distances. This alone has contributed to more and more plants adopting the area maintenance structure in recent years. Obviously, production likes this concept because it brings maintenance resources close at hand; they get to know the foreman and the tradesmen on a first-name basis. Similarly, the maintenance people get acquainted with their production counterparts. Obviously, this has a positive effect on their working relationships.

Advantages of Area Maintenance

- faster response time due to less travel
- greater familiarity with specific machines
- promotes team spirit and personal accountability
- shortens the lines of communication
- easier to coordinate multicraft jobs
- fewer lines-of-demarcation problems
- greater dedication to scheduled routine maintenance
- close proximity improves customer relationships

Disadvantages of Area Maintenance

- less communication outside of the area
- area focus reduces sensitivity to plant-wide priorities
- limits skills development to area machines
- frequent overstaffing and underutilization
- much greater impact of absence
- more difficult to administer balancing of overtime hours
- requires duplication of special tools and equipment
- promotes overstocking of parts and materials

Flexible Area Maintenance

This later version of area maintenance is (for some plants) arguably the best approach to operational maintenance structure, having some of the advantages of both centralized and area maintenance.

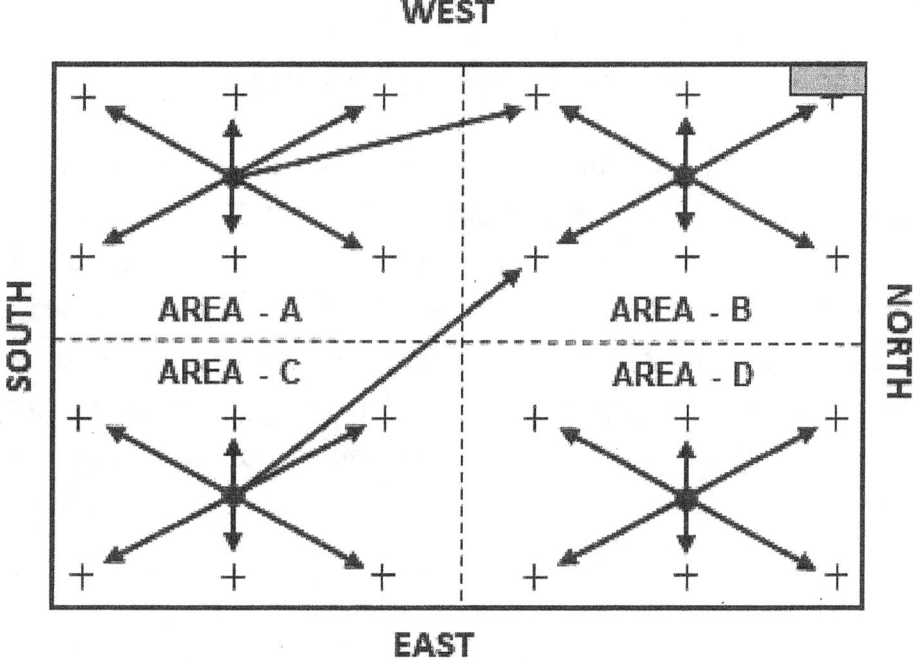

Note that in the previous drawing, both area A and area C have assigned resources out of their area to assist in area B. This demonstrates the ability to manage an area and at the same time be responsive to plant-wide priorities. This provides the most effective and efficient responses to machine downtime.

Advantages of Flexible Area Maintenance

- benefits of both area and centralized concepts
- both area and plant-wide priorities managed simultaneously
- promotes a wider range of craft skills and experience
- more efficient utilization of skilled maintenance resources
- ensures resources being used in the area of greatest need
- reduces response time and related machine downtime
- reduces impact of absence in the benefitting area

Disadvantages of Area Maintenance

- resistance to assignments outside of the area
- tendency of area foremen to hide resources
- the area's own work gets delayed
- workload is overstated to get extra help
- need for additional controls and training

Integrated Production Teams

The use of integrated production teams consisting of machine operators, setup, quality control, inspection, materials management, maintenance, etc., has proved to be highly successful in some cases, and in others not so much. The areas within the dotted lines in the drawing below could represent the location of the teams. Each team could represent a product, a process, a machine, or a group of machines. As with other organizational structures, there are advantages and disadvantages, which are presented in the context of how they affect plant maintenance operations.

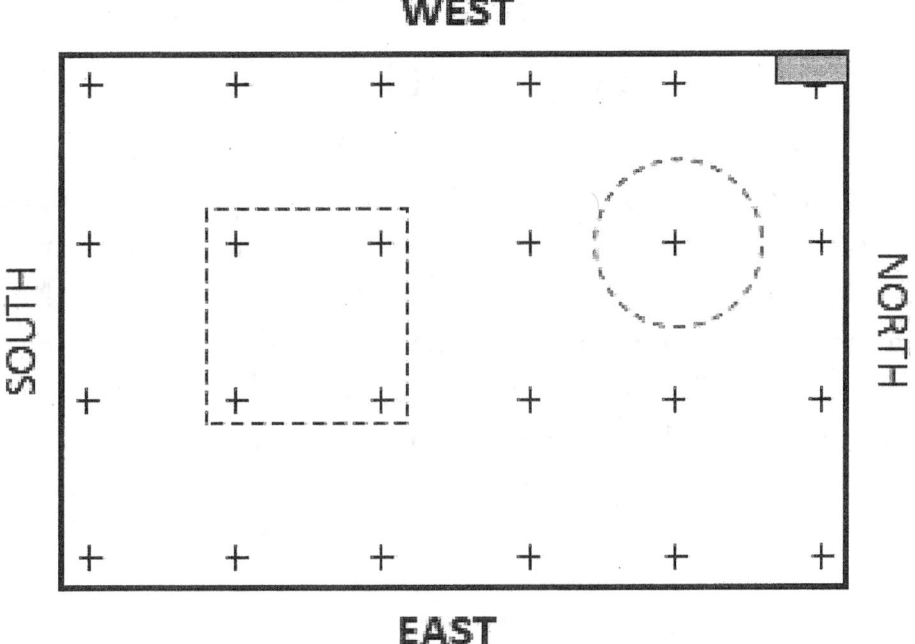

Advantages of Integrated Production Teams

- customers responsible for their own maintenance
- maintenance supported by production team
- travel minimized

Disadvantages of Integrated Production Teams

- limits career development of maintenance employees
- lack of technical leadership and support
- difficult to implement common maintenance systems
- lack of strategic maintenance planning
- difficult to assess maintenance effectiveness
- maintenance employees commonly underutilized

This information should help management consider all of the advantages and disadvantages of structuring maintenance resources when making their own structuring decisions.

Chapter 11
Management Reports

The evolution of computers has played a leading role in bringing about what has been called the *information age.* Data is now seen as a major asset, joining other major organizational assets like human resources, physical assets, and intellectual assets. A computer database provides the ability to select needed data and present it in different formats. For example, a machine uptime report, using the same data, might be displayed in the following formats:

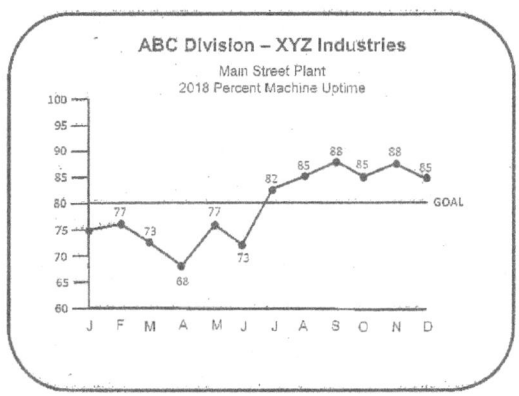

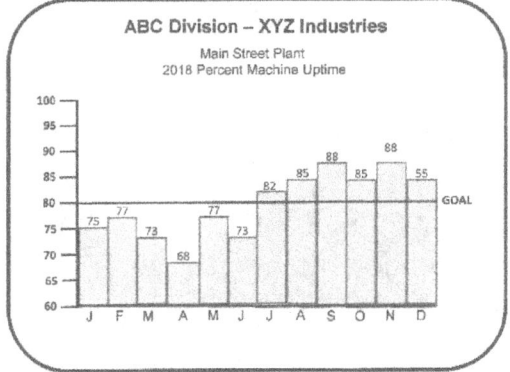

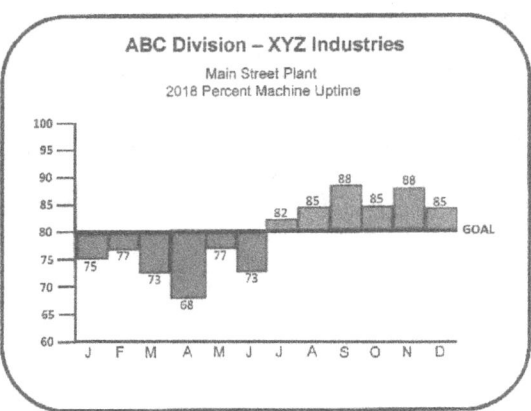

As long as the information is in the database, maintenance department workers get to pick and choose what they want to see, and in the format of their choice. For example, maintenance expenditures could be displayed as a pie chart:

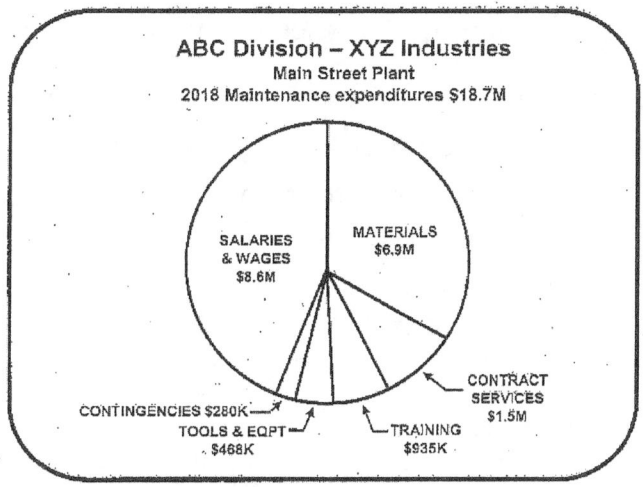

On the other hand, *skilled-trade employee age versus years of service* demographics might be displayed using an X-Y format:

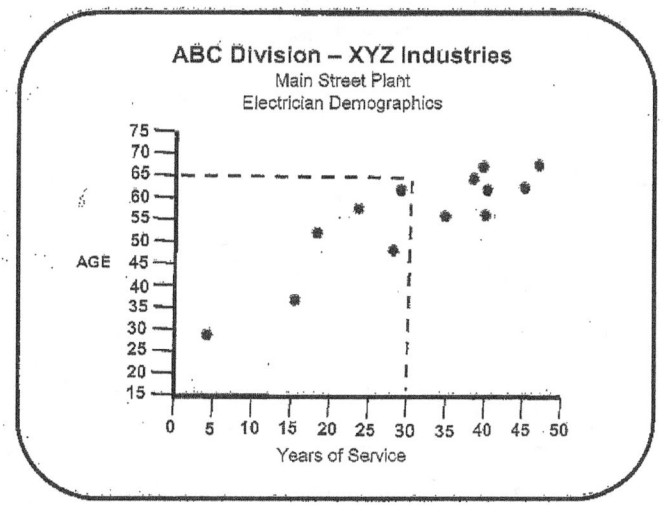

The maintenance department has been operating for so long with limited information, having it will be a new experience that requires some getting used to. It will quickly become evident that managing by the numbers is a far more effective way of handling the business of plant maintenance. Managers' access to timely information not only allows them to see where actions may need to be taken, it also lets them see the results of those actions. This is the gold standard for management reports.

Information can also be presented in traditional tabular form. For example, how are maintenance resources being utilized? In most plants, that information usually comes only in the form of labor and materials. In some cases, labor and materials are broken down into scheduled and unscheduled activities. Imagine being able to see the information below, especially at budgeting time.

ABC Division – XYZ Industries

Main Street Plant

2018 Maintenance Labor Summary

Maintenance labor charged by craft is expressed in Man/Years

DEPARTMENT	E	M/R	T/R	PF	MW	W	P	O	L
UNSCHEDULED MAINTENANCE									
111	0.80	0.95	0.24	1.07	0.37	0.04	0.00	0.02	0.33
222	1.73	1.21	0.08	1.03	0.51	0.00	0.00	0.03	0.12
333	1.48	1.76	0.79	0.96	0.76	0.03	0.00	0.00	0.36
444	1.81	0.56	0.80	0.49	0.43	0.00	0.00	0.01	0.28
555	1.43	2.51	0.85	0.90	0.91	0.02	0.00	0.01	0.23
TOTAL	7.25	6.99	2.76	4.45	2.98	0.09	0.00	0.07	1.32
SCHEDULED PREVENTIVE MAINTENANCE									
111	0.97	0.21	0.20	0.26	0.43	0.00	0.00	0.44	0.24
222	0.67	0.43	0.17	0.66	0.32	0.00	0.00	0.70	0.19
333	1.34	0.94	0.29	1.01	0.83	0.00	0.00	0.45	0.86
444	0.88	1.10	0.61	1.24	1.09	0.00	0.00	0.43	0.43
555	1.33	2.03	0.41	0.89	0.14	0.00	0.00	0.91	0.96
TOTAL	5.19	4.71	1.68	4.06	2.81	0.00	0.00	2.93	2.68
WORK ORDERS									
111	0.09	0.05	0.21	0.04	1.06	0.48	0.07	0.00	0.00
222	0.02	0.09	0.11	0.17	0.89	0.58	0.88	0.00	0.00
333	0.16	0.08	0.03	0.11	0.77	0.32	0.47	0.00	0.00
444	0.13	0.06	0.09	0.15	1.14	0.61	0.39	0.00	0.00
555	0.16	0.02	0.12	0.02	1.35	0.92	1.19	0.00	0.00
TOTAL	0.56	0.30	0.56	0.49	5.21	2.91	3.00	0.00	0.00
GRAND TOTAL	13.00	12.00	5.00	9.00	11.00	3.00	3.00	3.00	4.00
HEADCOUNT	13	12	5	9	11	3	3	3	4
VARIANCE	0	0	0	0	0	0	0	0	0

If the maintenance department is going to manage effectively, it needs the right kinds of data to manage by the numbers. It also needs the data presented in forms that are easily understood. What has been presented here is just a small sampling of the ways in which data can be presented to management.

Chapter 12
Core Business Focus

There are successful companies that have had to learn the hard way not to stray too far from their core business. Some have extended themselves into products and services in which they have no prior experience and later found to their dismay that the added activities were distracting them from their core business. What many if not most medium and large discrete manufacturing plants have yet to learn is that their maintenance departments may be involved in activities outside of their core business, which is to preserve the plant's physical assets. In fact, this is the sole purpose of plant maintenance!

Although there are numerous examples of how plants have employed their limited skilled resources outside of the boundaries of preserving physical plant assets, the chief culprit is minor construction and project work. Correcting this problem is made more difficult by the fact that this type of work may have been going on for so long that it's being mistakenly viewed as a part of the role of plant maintenance, with no awareness of how this adversely impacts the performance of the larger business enterprise.

This problem wasn't so evident in the era prior to World War II. However, events that have taken place in the years following the war have contributed to compromising the core business of plant maintenance through vertical integration within the maintenance department. Immediately after WWII, there were many with maintenance experience in the military who returned to their jobs in manufacturing plants, while others returned to jobs in the construction industry. Later, during the years of economic expansion, shortages developed for skilled workers in plant maintenance. Over time, construction workers (those having the necessary skills) began to fill some of those job openings, recognizing that plant maintenance was offering them greater job security, a shelter from working in inclement weather, and many other benefits.

Because plant maintenance had historically promoted from within, it was only natural that some of these former construction workers were eventually promoted into positions of responsibility, such as foremen, second-level general foremen, and third-level managers. As the responsibilities and influence of these former construction workers increased, the amount of construction-type work being performed by the maintenance department also increased. Although a limited amount of minor construction work had been done in the past, a greater focus was now being placed on construction work. Not only were more and more minor construction jobs being performed, much larger and more time-consuming projects were involved. This increasing level of construction work had nothing to do with maintaining the plants' machinery, equipment and other physical assets. In fact, it was working to the detriment of that mission.

Unless the legitimacy of these activities is questioned, they will continue, because the maintenance department's customers are only too happy to develop their laundry list of wants and needs. This trend has led many plants to reorganize the maintenance department into three separate product lines: production support, plant facilities, and construction. Increased levels of minor construction and project work inexorably leads to higher costs via adding head count, tools, equipment, and supplies. It has been through this evolving chain of events over the years that plant maintenance has inadvertently become less cost efficient and has, in the process, added to the cost of production.

The increase in construction activities has taken some very strange twists and turns indeed. Consider how something as common as a reduction in production volume can, in many cases, lead to more construction-type work. For example, if production volume is down to a level that justifies reducing head count in the maintenance department, there will likely be strong resistance. First of all, the maintenance manager doesn't want to lose valuable skilled workers, who may not return when times get better. In many cases, a reduction of maintenance employees (even when justified) will never take place due in part to the manager's concerns, in concert with a strong union position against the reductions. The union may argue that work is being done by outside contractors that could be performed by represented employees, even though the type of work involved has never before been performed in-house. It's also suggested that the work can be done in-house for less than contracting it outside. This is pure voodoo economics and very misleading.

This "we can do it cheaper" issue has been raised on many occasions. When those estimates have been thoroughly reviewed by the accounting department, the in-house costs are almost always higher because of internal costs that are never included. The legacy costs of the contractor are minimal compared to those in a manufacturing plant. Secondly, the added work may directly or indirectly result in overtime that was not included in the estimate. The contractor avoids overtime by using temporary help. In addition, the contractor will utilize major labor-saving machinery and equipment not normally found in the maintenance department.

Agreeing to do the work in-house may satisfy both the union and the maintenance manager's concerns, but what about the future effect? Once the work has been performed by a represented employee, it becomes represented work by past practice. When volumes rise again, the maintenance department will be busier than ever due to the added workload. This in turn leads to more of what have become maintenance department staples—overtime or adding head count—all because of engaging in non-core maintenance activities.

Perhaps the greatest adverse effect of construction work is that production support operations (the core business of plant maintenance) must compete with construction activities for both priorities and the plant's limited skilled resources. Few if any manufacturing plants have ever resolved this issue. Even when procedures are in place to borrow people from construction to support production during peak demand, an infinite number of variables will almost always delay maintenance response to machine failures and result in increased machine downtime.

Machine downtime increases when the employees performing construction work are temporarily assigned to help in supporting production. Employees who normally perform construction work will always take longer to diagnose and correct machine problems simply because they lack the familiarity and experience of those who work on these types of machine problems every day. Yet another negative aspect of having people working in construction activities is that if they are in the same hours-balancing/overtime-equalization group as those supporting production, this can become a very costly overtime problem.

For example, department 300, a high-tech highly mechanized manufacturing area, will be working Saturday. The problem is that the maintenance people supporting that area are high on the hours-equalization list. The people who should be asked first to work overtime have little or no experience with this kind of equipment. The manager obviously must bring in the experienced people. In addition, he must find some kind of Saturday work for the people who are low on the list, even though overtime isn't justified; it's either that or be faced with possible contractual violations and end up paying overtime for hours not worked. Bringing in more people on overtime than are needed just to get the right people on the job is altogether too common, especially in large manufacturing plants. These kinds of problems increase the cost of doing business.

There is sound advice for those companies whose maintenance departments have never become involved in construction-type activities: *don't start*. There is also advice for those companies whose maintenance departments are currently engaged in construction-type activities: *get out of the business!* While that may seem impractical, it can be done over time by refocusing the plant's skilled resources on the core business of plant maintenance. Doing everything that should be done to support production operations would probably require all of the plant's skilled resources. If everyone is focused on preserving physical plant assets, there will be no time for construction activities.

If you think about it, focusing all of a plant's skilled resources on the core business of plant maintenance will first of all reduce machine downtime, and its associated production costs and losses, to a minimum level, ultimately reducing the requirement for maintenance. This in turn allows for managed adjustments in maintenance headcount until optimum levels are reached.

Chapter 13
Worker Productivity
Why Good People May Perform Poorly

A 2009 article in *USA Today* focused on lost worker productivity in the United States. It was estimated that the average worker is nonproductive 1.7 hours each day. It further stated that this loss in productivity is costing American businesses billions of dollars each year. If lost productivity had been focused solely on plant maintenance workers rather than workers in general, the level of lost productivity would have been greater. Nevertheless, it shouldn't be assumed that maintenance workers are the primary cause of lost productivity, because they are not. This isn't a worker problem; it's a management problem.

The maintenance manager's attention is focused primarily on two key issues: 1) keeping the plant machinery, equipment, and facilities operating, and 2) controlling the maintenance budget. Consequently, worker productivity doesn't usually get a lot of attention. One of the reasons for this may be the subtle way in which productivity can degrade so slowly over time that it appears on the scene virtually unnoticed. If top-level managers were aware of the extent to which maintenance-worker productivity affects overall plant performance, it would get a lot more attention at all levels of management—and sooner rather than later.

Four Dimensions of Worker Productivity

There are significant differences in the management of direct labor as opposed to indirect labor, such as plant maintenance. For example, the productivity of production workers is engineered into the manufacturing process. Each worker is performing specific repetitive assembly, machine, or process tasks when the production part is at their work position. The time for performing each task is determined by machine cycles, transfer rates, sampling, or some other standard engineering method. The productivity of direct labor is relatively stable, unless there is an interruption or constraint somewhere in the manufacturing process. If problems do occur, supervisory assistance is normally close enough to the scene to deal with them.

The plant maintenance work environment is far different. There may be little direct supervisory contact during the work shift. The work is mostly nonrepetitive, and work tasks can vary significantly from job to job. The work site may be anywhere in the plant. It changes frequently, and worker productivity is subject to a host of on-the-job work constraints that have a negative impact on worker effectiveness, efficiency, utilization, and motivation. The challenge is to better manage these unique dimensions of worker productivity.

Effectiveness

Effectiveness relates to how well workers perform the job tasks to which they are assigned. For example, does an electrician understand complex electrical-wiring schematics and drawings? Does the employee follow safety standards and safe practices on the job? Can the employee diagnose and correct electrical problems without assistance? Is work performed correctly the first time, every time? Are tools and leftover materials removed, leaving the jobsite clean and orderly after the work is completed?

These and other indicators measure a worker's effectiveness. Maintenance workers have demonstrated again and again that when they have the necessary knowledge, skills, experience, tools, materials, and other logistical support, they will be effective in the workplace.

Efficiency

Efficiency relates to how much or how little waste is involved in performing on-the-job tasks. Maintenance workers are daily confronted with (preventable) work constraints, roadblocks, barriers, and delays, such as waiting for something or someone, excessive travel to and from the jobsite, lack of adequate job information, and searching for parts, tools, and materials. If a two-hour job takes four hours, don't automatically assume the worker is at fault. In too many cases, the fault lies with management.

On-the-job work constraints can more often than not be traced to dysfunctional organizational structure, systems, processes, policies, procedures, and practices. These are the management tools that have been put into place to control the operations of the business enterprise. When these tools become dysfunctional, as they often do over time due to environmental changes, only management can correct the situation. It's either that or continue the preventable waste.

Utilization

Utilization relates to how much of the time during the work shift employees are assigned to and engaged in productive work. There is little disagreement about the need for every maintenance worker to have a job assignment; that doesn't mean everybody will have one. It's not uncommon for maintenance workers to be without an active job assignment, sometimes for extended time periods. This situation probably occurs more frequently when workers are foreman-dispatched. When an employee completes a job assignment, the maintenance foreman may not be available to make another assignment. The foreman may be out in the plant providing supervisory assistance to other employees, or chasing parts, or expediting purchase order approvals, or conferring with production on the status of a down machine, or attending a planning meeting.

Some have tried to resolve this problem by giving workers multiple job assignments, to be performed in a certain order. The problem with the job-queue approach is that conditions in a manufacturing plant are constantly changing. After a worker completes one job assignment, the next in the queue may not now be the job the employee should be assigned to, because other more important work has since emerged and priorities have changed.

Motivation

Motivation represents another important factor in worker productivity. Even though first-line foremen may never have been formally trained to deal with motivational issues, that doesn't mean they can't have a positive influence on employee motivation just by using a little bit of common sense. One of the commonsense things a foreman needs to be aware of is the *can't-don't-won't* worker syndrome.

Workers *can't* be fully productive if they lack:

- knowledge
- skills
- experience

- tools
- spare parts and materials
- adequate job information
- up-to-date prints and drawings
- effective communication processes
- timely on-the-job assistance
- logistical support

Workers *don't* become fully productive because of adverse consequences:

- peer pressure
- unfair workload distribution

Workers *won't* be fully productive if they lack:

- motivation

It becomes obvious why workers *can't* be fully productive when they lack the things that are required and necessary for them to be fully productive. Why workers *don't* become fully productive may not always be so obvious.

For example, if an employee is fully productive while others are not, it can expose that employee to peer pressure. The productive employee may be accused of making everyone else look bad, or apple-polishing the boss, or looking for special treatment. At the same time, being fully productive may result in that employee having to bear a greater, disproportionate amount of the work, while others don't do their fair share. This can result in employees who have been very productive in the past feeling like they are being taken advantage of and becoming far less productive in the future.

If some workers aren't doing their fair share, the manager needs to deal with that quickly, because it's unfair to all of the other workers involved. Remember, if management doesn't take action to deal with poor performers, that can lead to the much larger problem of workers feeling that management doesn't really care about everyone being productive. If management doesn't care, why should they?

Finally, there are those employees who won't be fully productive simply because they lack the motivation. When confronted with an unmotivated employee, a manager should approach the problem in a positive manner. There may be good reasons for the behavior.

Consider such employees' past performance. Has their behavior changed over time? How do they interact with others in the workplace? What kind of person are they outside of the plant? Are they engaged in community service activities? Do they hold a public office of some kind? Have they received achievement awards? Have they pursued higher education? Does the employee have health problems? What about family? There are so many things that could influence performance on the job.

The problem may also be that these employees are not being challenged. They may be far more capable than they are allowed to be, which can easily demotivate a highly capable individual. In such cases, it's possible to transform these employees into highly productive workers just by giving them some responsibility or a challenge that offers an opportunity to better demonstrate their capabilities. If that's successful, give them recognition. A sincere thank you for a job well done may be all it takes.

In spite of management's best efforts, there are a few individuals who will not respond to a positive approach. They will continue to perform poorly. In such cases, there are various negative rewards that can have a motivating influence. One technique that has proven successful is to assign such employees to a captive job that keeps them in one specific work location, possibly a parts or repair crib, or some other location where their activities can be more closely supervised. It's possible that a captive job assignment will motivate the employee to become more productive simply to escape the confining work environment. However, a word of caution: never apply negative rewards until the positive approach has been tried first and failed.

Organizational Culture

Organizational culture also influences worker productivity. Organizational culture can be thought of as the organizational attributes that help form employees' perception of the organization, management, coworkers, their work, and themselves.

For example, trust and respect (or the absence thereof) are attributes of organizational culture that can affect productivity. Management wants the trust and respect of its employees no less than the employees want the trust and respect of management. If trust and respect become an issue, nature offers an important perspective: whatever people plant is exactly what they are going to reap. You cannot plant wheat in your field and expect to harvest corn.

What are you planting, Mr. Manager? If you want to be trusted and respected, you must trust and respect others. Trust and respect issues don't just somehow disappear on their own; something has to take place first to make that happen. The ball is in management's court. Managers are in the best position to take that essential first step. Give to others what you expect from others.

Needs and Goals

In addition to all of the other things that influence employee productivity, needs and goals play an important role and affect both the company and its employees. Consider the following statements:

- Employees ultimately determine the success or failure of the organization.
- Employees associate themselves with the organization to meet their own needs and goals.
- To the employee, their needs and goals supersede those of the organization.
- Dysfunction occurs when the needs and goals of the organization and the needs and goals of its employees conflict.
- Success results when the needs and goals of the organization and its employees are both satisfied.The above statements were found by an Electronic Data Systems Employee on his desk in Troy, Michigan, in 1991.

Recognizing that the company's success is dependent upon its employees, and that the success of the employees is dependent upon the company, this should be a no-brainer. Make it a win-win situation.

Chapter 14
Operator Machine Care

Among the most important issues in any manufacturing plant are employee safety and avoiding possible damage or loss to the plant or its physical assets. Usually, there is extra emphasis given to physical assets like machinery and equipment. During preproduction qualification, every part of a machine should have been inspected to ensure it is hazard-free for the operator or others who may be at or around the machine while it is in operation.

Normally, potential safety hazards have already been considered by the machine-builder, who has installed safety gates, guards, shields, and optical devices that detect the presence of objects inside an unsafe area and keep the machine from cycling. These and other safety devices are some of the common methods being used to protect the operator and others from injury, but remember: safety is the plant's responsibility. It cannot be left to someone else.

While these safety devices are very reliable, they are not failure-proof, nor are they maintenance-free. They must be inspected periodically not only for safety's sake but also to comply with OSHA and insurance requirements. The question is, how often must they be inspected? Some might say, "As often as it takes to comply with requirements." That's okay, but it may not be often enough to prevent a serious injury or even death. It might be better to say, "Do everything necessary to prevent an accident."

Here is where the machine operator can provide an extra margin of safety. Instead of relying only on monthly inspections by the maintenance department, as most do, why not rely on the person who is at the machine every day, all day? In most cases, a brief inspection would require only minutes at the start of the shift each day.

For example, how long does it take the machine operator to make a single circuit around a machine to check that all machine guards are in place and undamaged? This task is easier when the machine guards are made highly visible by painting them the diagonal-striped black and yellow safety colors.

Unfortunately, the role of machine operators in the maintenance of their machines has at best been underutilized, or at worst, never even considered. The machine operator could, and should, also become the first line of defense in the prevention of machine failures. The operator is at or in close proximity to the machine every minute of its operation and so becomes sensitive to the unique personality-like traits, nuances, and quirks (if you will) that are inherent to individual machines. In this sense, operators know more about their machine and its operation than anyone else. This knowledge can provide invaluable assistance to the maintenance technician trying to diagnose a malfunction. Allowing the maintenance technician and the machine operator to work together can lead to quicker fixes and less machine downtime.

If this is not current practice, make it standard operating procedure. When there is a machine failure, the machine operator remains at the machine to provide any nontechnical assistance that might be needed in restoring the machine to production. Obviously, this is predicated on the fact that the operator isn't immediately needed for production elsewhere and that such assistance will not extend beyond the shift on which the failure occurred.

If there are contractual labor agreements covering skilled maintenance employees, additional contract language may be required to avoid possible conflicts. A straw-man model for such language could be as follows:

> In case of a machine failure, the machine operator will remain at the machine to provide any nontechnical assistance that can be safely performed under the direction of the maintenance technician. Examples might include things such as cycling the machine, holding a light or tool, communicating with others, obtaining parts or materials, providing a second-person safety presence, and calling for additional technical assistance.

On-the-job operator assistance will help ensure that the limited skilled resources of the maintenance department are available for the more important work for which they alone are technically qualified. They will not be distracted by simple nontechnical tasks that anyone can perform under the direction of the maintenance technician.

There is another way in which the involvement of machine operators can help to prevent machine failures. Machine tool technology has evolved over the past fifty years from mostly single-purpose cutting, punching, forming, and joining to highly sophisticated, computer-controlled, multipurpose machine cells. As machines have changed, the role of the operator has also changed. In the past, the operator may have done little more than feed parts and push palm buttons to initiate machine cycles. Today, operators may also perform their own setup, making computer program changes to modify both the function and operating parameters of the machine. In some cases, machine operators perform their own parts inspection.

The majority of these changes have come about through the use of very expensive technology with significantly higher maintenance costs. This is where machine operators can make a big difference by becoming involved in the preventive maintenance of their own machines. It doesn't matter whether production machines are high-tech or not; the machine operator can become a key element of a plant's preventive maintenance activities, and it's virtually cost-free.

Think about it: operators are always at their machine; they know more about its operation than anyone else; and they were born with the sensory mechanisms of sight, sound, touch, smell, and taste. These natural powers enable them to receive information about what's going on around them. What a waste if these powers are not fully used on the job! They can see when something isn't working properly. They can hear abnormal sounds. They can feel vibrations. They can smell and even taste unusual odors. More than that, they have logic, reasoning, and response capabilities exceeding that of computers.

All that's needed is a little training to make them sensitive to the early-warning signs of a machine problem, and they can become first responders in the prevention of machine failure. When machine operators are made to feel a sense of ownership of their machine and a sense of responsibility for maintaining it, their involvement becomes much easier.

As machines have become more complex, so has the role of the operator. This creates the need for formal written operating procedures (and related training) for each type of machine. In addition, start-up procedures should be documented and a printed log used to confirm that each step has been completed. This is not as complicated or time-consuming as it might appear at first glance. In fact, with one machine, it might take only twenty to thirty seconds to complete the log. For another very complex machine, it might take three to five minutes or more. This is a very small price to pay for the best kind of safety and liability insurance.

If this would prevent just one employee injury or one costly machine failure, it would be well worth the time and effort. Nobody wants to be faced with an injured employee or worse. However, it would be important to have the training log signed by the employee, indicating that the employee was trained in the proper operation of the machine, including safety procedures. This plus the daily start-up log initialed by the operator indicating that those procedures had been followed would ensure that everything possible had been, and was being, done to prevent a mishap. It would provide hard evidence of the plant's safety activities when OSHA comes knocking on the door.

There are other things that should be a part of an owner-operator's responsibility— for example, all daily lubrication, plus inspecting the operation of automatic lubricators and all fluid reservoir levels. The Japanese have proven the link between dirty machines and machine malfunction; beyond that, everyone likes working in a clean environment. Owner-operators should be responsible (as much as possible) for the cleanliness of their own workstation. When more general machine-cleaning is being scheduled, owner-operators should be offered the off-shift and/or overtime opportunities for cleaning their own machines.

There is great diversity across industry in the various manufacturing processes and types of machines being used, and the role of the machine operator can be just as different. The information being presented here is not a one-size-fits-all solution. However, the ideas presented can inspire thoughtfulness that can make the operators' environment a safer place to work and, beyond that, utilize their invaluable talents and skills in the preventive maintenance of their own machines. That's a common-sense proposition for every plant.

Chapter 15
High-Tech Manufacturing and Plant Maintenance

One major area of increased maintenance cost is seen with the greater use of advanced high-tech production machinery and equipment. These assets are more expensive and much more complex. This creates a whole new set of maintenance issues.

First of all, additional time and training is needed to acquire the knowledge and skills necessary to maintain this type of machinery and equipment, and proper training is expensive. Not only is technical training expensive, the time required to train (and at the same time keep plant machinery and equipment operating) can be very difficult to manage. As the level of complexity increases, so does the probability that some employees, even though they have been trained, may never achieve the required level of technical competence. Training is one thing, but learning is influenced by individual capabilities and the desire to learn.

Another cost factor associated with high-tech machines is expensive replacement parts and components. In many cases, these replacement parts must be kept in-house for possible immediate use. Idle in-house spare-parts inventory becomes yet another form of added costs.

In addition, these marvels of modern machine technology are subject to a broader range of failure, in which both the frequency and duration of downtime can be greater due to the additional time required to diagnose problems with integrated solid-state electronics and mechanical, hydraulic, and pneumatic control systems. Additional diagnostic tools and equipment will also be needed by the maintenance technicians performing hands-on repairs.

While high-tech machinery and equipment can improve product quality and reduce direct labor costs, at the same time, they can add significantly to the cost of maintenance. The following insights may help explain why maintenance costs are so high and why these costs are continuing to increase.

Why Do Machines Fail?

There are many reasons why machines fail. However, there are some that occur again and again across industry, and most of them are preventable. The following listing is in no specific order or significance. Rather, the items have been listed randomly:

- running production prior to fully qualifying the machine
- running production prior to fully qualifying the tooling
- complexity of electronic, hydraulic, and pneumatic control systems
- poorly designed machinery and equipment
- improper application of machines to process
- improper operation/operator error
- running machines at excessive speeds and feeds
- ignoring incipient problems
- run-it-until-it-breaks attitudes
- failure to properly monitor machines during start-up/operation
- excessive platen pressure to correct die wear
- improper machine setup
- wrecks at transfer points
- variability of raw materials/out of specs
- normal/abnormal wear of parts and components
- tooling breakage, wear, and changeover
- utility or process supply failures
- environmental factors such as dust, dirt, corrosion, etc.
- deliberate actions that cause failures
- lack of consistent, quality PM/routine maintenance
- poor-quality replacement parts and components
- poor-quality previous repairs
- improper lubrication—wrong/too much/too little
- inadequate operator training and experience
- inadequate communication

Maintenance, Machine Malfunction, and Murphy's Law

If you think this sort of thing doesn't happen in your plant, think again:

- A malfunction will go undetected until the machine or production parts have been damaged.
- When a malfunction has been detected, there will be a long delay for maintenance assistance because everyone will think that somebody else has called for maintenance.
- When it is finally realized that maintenance hasn't been notified, all of the telephone lines in the maintenance department will be busy.
- When a phone line to the maintenance department becomes available, the foreman will be out in the plant ... somewhere.
- When the foreman is finally located, nobody is available because everyone else is currently assigned to other high-priority jobs.
- When a repairman becomes available and arrives at the machine, the operator isn't there to discuss the problem.
- When the operator arrives and the problem has been diagnosed (say, a broken compression spring), the parts catalog needed to identify the part number can't be found.
- When the catalog is located and the part has been identified, the part is out of stock.
- When the new part is ordered and finally delivered to the receiving dock, there is a delay because nobody has yet notified the maintenance department that the part has arrived.
- When the part is obtained by maintenance, the special compression tool needed to install it can't be found.
- When the tool is finally located, there is a delay because it requires two people to install the spring and nobody is available to assist.
- When assistance becomes available and the installation of the spring has been completed, there is a delay because the operator isn't there for a start-up and tryout.
- When the operator finally arrives and the machine is declared ready to go, guess what? They can't run production because they have run out of raw material.

Anatomy of Machine Downtime

There are several reasons for machine downtime unrelated to maintenance. The following common contributors to machine downtime involve different areas of responsibility:

- production process problems
- quality issues
- material problems
- operator availability
- safety issues
- machine idling
- power failures
- accidents
- parts jammed
- reduced operating speeds
- machine malfunctions

Regardless of the reasons for machine downtime, the consequences are the same: production costs and losses. Stated differently, downtime increases the cost of production.

The focus here is solely on maintenance-related machine downtime. To most people, downtime is just downtime. They don't consider the factors that contribute to the duration of machine downtime. Unscheduled machine downtime must be avoided whenever possible, but when it does occur, it's the duration of downtime that becomes the all-important issue. The following series of events takes place from the time a machine malfunctions until the problem is corrected and production on the machine is resumed.

- A machine problem occurs.
- Production on the machine stops.
- Maintenance assistance is requested.
- Maintenance resources are assigned.
- Diagnostics are performed to identify the problem.
- Repair parts, tools, etc. are obtained.

- The problem is corrected.
- Normal production is resumed.

It's during and between these events that things can happen to extend the duration of machine downtime. Obviously, one objective is to identify the primary cause of the machine failure and then take whatever steps are necessary to prevent another failure from the same cause.

There is another important objective that may be missed in the heat of a busy day. What about the delays and roadblocks that make getting the machine up and running again take longer than it should? If nothing is done to prevent it, those things will continue to happen again and again. Don't miss out on another great opportunity for preventive maintenance.

Chapter 16

Machine History

A Super-Tool for Maintenance Managers

In the days before computers were used in the maintenance department, whatever history was documented about the work performed on machinery and equipment was kept on handwritten paper and stored in file cabinets. These paper records were maintained because of the gut feeling that the information might become invaluable in the future to help solve machine problems, identify faulty repair parts, justify machine modifications, etc. The usefulness of paper records was impaired because of poor handwriting, dirt, smudges, and fading. As time passed, one filing cabinet became a row of filing cabinets that occupied too much office space.

The actual use of this information was limited because of the time required to get useful information. For instance, it was virtually impossible to detect whether the same part number was failing on multiple machines. Even if the information was available, it became impractical due to the time required. The coming of computers into the maintenance department solved that problem … or did it? If the same kind of information was being put into a computer system, in the same format as before, the same time-consuming search problems would still be there.

There was a humorous piece widely circulated in the 1980s and 1990s relating to the application of computers. It's too bad it's anonymous, because the author deserves credit for what is really a brilliant observation: "If you computerize a system that is all fouled up … it now fouls up automatically, faster, more consistently, and a lot more efficiently."

You can't use computers like a pencil and paper—that is, if you want to search for information. For this purpose, computers use symbols or short codes to represent data. Creating a system of machine codes makes it possible to quickly retrieve a wide variety of machine-history information. Unfortunately, there is no universal coding system for machine history. This is due in part to the different kinds of machines that are in different plant locations. The codes must be developed to fit the machines in each individual plant.

Experience has shown that involving the people in each plant in the creation of their own codes will virtually guarantee a coding system that meets their unique requirements and one that can be used successfully. The thought process when developing machine codes is what information is needed to create machine histories that will provide a record of the following:

- maintenance work performed
- machine problems and problem machines
- repetitive machine problems
- opportunities for preventing machine failure
- data for machine diagnostics
- data for machine reengineering
- justification for machine modifications/replacement
- identification of faulty machine parts and components
- documenting the effects of management actions

For the purpose of facilitating the development of an in-house machine-coding system, it's sometimes helpful to see examples. These should not be seen as something to copy but rather as ideas or thought-starters. The codes must fit the machines at each plant location. In the following examples, three types of codes are used, each serving a unique purpose: action codes, component codes, and cause codes.

Action codes describe an action taken, such as *replace*. Component codes describe the component involved, such as *bearing*. The action code is combined with the component code: *replace bearing*. Then a cause code is added: *lack of lubrication*. The three codes together—*replace bearing/lack of lubrication*—would be entered into the computer as three sets of numbers—26 037 .31

Retrievable information includes the following:

- date of occurrence
- machine name, number, and department
- bearing part number and cost
- craft skills required
- labor hours and cost
- total job cost

This information is added to existing data, which then can be quickly retrieved from the computer and printed as a hard copy report. The following is an example of a coding system.

Action Codes

01 Add
02 Adjust
03 Assemble
04 Assist
05 Calibrate
06 Change
07 Clean
08 Connect
09 Cut/Cutting
10 Disconnect
11 Fabricate
12 Fill/Refill
13 Hookup
14 Inspect
15 Install
16 Level
17 Light/Relight
18 Loosen
19 Lubricate
20 Modify
21 Move/Relocate
22 Program
23 Realign
24 Rebuild
25 Repair
26 Replace
27 Reset
28 Start-up
29 Tighten
30 Troubleshoot
31 Weld
32 Wire/Rewire
33 Other

Component Codes

001 Air Conditioner
004 Alarm System
007 Amplifier
010 Anvil
013 Arbor
016 Arm
019 Armature
022 Atmosphere/Gas
025 Axle
028 Ballast
031 Base/Frame
034 Battery
037 Bearing/Bushing
040 Belt/Conveyor
043 Belt/V
046 Belt/Vari-speed
049 Blade/s
052 Blower/Exhaust
055 Bolster/s
058 Bolt/s
061 Bracket/s
064 Brake
067 Burner/Gas
070 Burner/Nozzle
073 Burner/Tile
076 Cam/Assembly
079 Chain
082 Chute/In
085 Chute/Out
088 Circuit/AC
091 Circuit/DC
094 Circuit Board
097 Circuit/Breaker
100 Clamp
103 Clock
106 Clutch
109 Coil/Cooling
112 Coil/Heating
115 Coil/Magnetic
118 Compressor
121 Computer
124 Computer/Monitor
127 Computer/Printer
130 Condenser
133 Container

136 Control/AC
139 Control/DC
142 Control/Hydraulic
145 Control/Mechanical
148 Control/Pneumatic
151 Control/Vacuum
154 Conveyer/Load
157 Conveyer/Scrap
160 Conveyer/Unload
163 Coolant
166 Cord
169 Coupling
172 Coupling/Drive
175 Counter
178 Counterbalance
181 Cover
184 Cradle
187 Crane
190 Crankshaft
193 Cutoff Mechanism
196 Cylinder/Hydraulic
199 Cylinder/Pneumatic
202 Diameter/Mechanism
205 Diaphragm
208 Door/Access
211 Drain
214 Drawbar
217 Drill Head
220 Drive/Vari-Speed
223 Drive/Constant Spd
226 Drum
229 Duct/Exhaust
232 Elevator
235 Enclosure

238 Fan/Circulating
241 Fan/Exhaust
244 Fasteners
247 Faucet
250 Feed/Mechanism
253 Feed/Vibratory
256 Filter
259 Firebrick
262 Flywheel
265 Fuse
268 Gasket
271 Gate
274 Gauge
277 Gear/s
280 Gearbox
283 Gear-Reducer
286 Generator
289 Gibs/Ways
292 Graph Paper
295 Guards/Shields
298 Heater
301 Heat Exchanger
304 Heat Set
307 Hinge
310 Hoist
313 Hook/s
316 Hopper
319 Hose
322 Humidifier
325 Hydraulic Oil
328 Indicator
331 Index Mechanism
334 Interlock/Safety
337 Jack-stand
340 Jaws
343 Lamp

346 Laser
349 Latch
352 Lights/Machine
355 Lights/Overhead
358 Line/Air
361 Line/Coolant
364 Line/Drain
367 Line/Gas
370 Line/Hydraulic
373 Line/Lubricant
376 Line/Vacuum
379 Line/Water
382 Lubricant
385 Lubricator/Manual
388 Lubricator/Auto
391 Meter
394 Motor/AC
397 Motor/DC
400 Motor-Starter
406 Motor/Synchronous
409 Oven
412 Photocell
415 Pilot-Light
418 Pilot-Nozzle
421 Pipe
424 Piston
427 Pitch Mechanism
430 Platen
433 Power Supply
436 Pre-Heater
439 Pump
442 Ram
445 Ram Assembly
448 Refractory
451 Regulator
454 Relay

457 Reservoir
460 Robot
463 Rollers
466 Roll-Feed
469 Scale
472 Screw/s
475 Segment Arm
478 Servomotor/diameter
481 Servomotor/Feed
484 Servomotor/Pitch
487 Shaft Assembly
490 Shear-pin
493 Sheave
496 Sheave/Vari-speed
499 Shot Media
502 Sight Glass
505 Slag & Sludge
508 Software
511 Solenoid
514 Sorter
517 Spindle
520 Spot Welder
523 Spring/s
526 Sprocket
529 Stapler
532 Straightener
535 Switch/Float
538 Switch/Flow
541 Switch/Foot
544 Switch/Limit
547 Switch/Photocell
550 Switch/Pressure
553 Switch/Proximity
556 Switch/Selector
559 Switch/Temperature
562 Switch/Misc.
565 Tachometer
568 Tail-stock
571 Tank
574 Tester
577 Thermocouple
580 Thermostat
583 Timer
586 Tool Changer
589 Tooling
592 Transducer
595 Transformer
598 Transmission
601 Trip Latch & Dog
604 Tubing
607 Turnbuckle
610 Turret
613 Valve/Air
616 Valve/Condensate
619 Valve/Gas
622 Valve/Hydraulic
625 Valve/Maxon
628 Valve/Motorized
631 Valve/Overflow
634 Valve/Steam
637 Valve/Vacuum
640 Valve/Water
643 Vane/s
646 Wear Plate
649 Wiper
652 Wire Feed
655 Wire Guide
658 Wire Roller

990 Other

Cause Codes

1 Accident	35 Misapplication of Eqpt.	71 Temperature/Low
03 Calibration		73 Scrap Buildup
05 Coolant Loss	37 Moisture	75 Speed or Cycles
07 Collision	39 Operator Error	77 Tension/Torque
09 Corrosion	41 Out of Sequence	79 Tightness
11 Contamination	43 Overload	81 Timing
13 Machine Design	45 Parts Jammed	83 Tool Breakage
15 Drain Clogged	47 Pilot Out	85 Vacuum/High
17 Filter Dirty	49 Plugged/Stopped	87 Vacuum/Low
19 Fuse Blown	51 Power Failure	89 Vibration
21 Imbalance	53 Power Surge	91 Voltage/High
23 Lighting	55 Pressure/High	93 Voltage/Low
25 Looseness	57 Pressure/Low	95 Waterflow
27 Production Material	59 Previous Repair	97 Wear/Abnormal
29 Lubrication/Wrong	61 Repair Part/Faulty	98 Wear/Normal
31 Lubrication/Out	63 Routine Service	
33 Misalignment	66 Rupture/Line/Hose	99 Unknown/Other
	69 Temperature/High	

After the coding system is complete, a small shirt-pocket-sized booklet can be constructed, similar to the one shown below, for reference when documenting the codes. Skilled trade employees are usually quick to memorize many of most commonly used codes, but the booklet is always handy for reference.

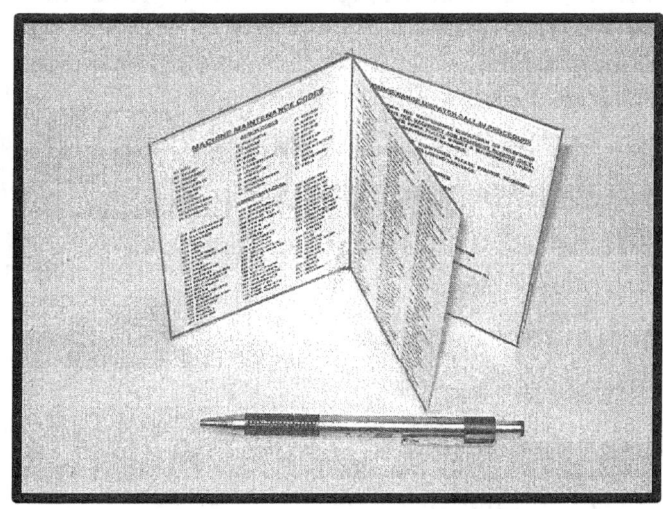

Machine history provides information to help busy managers better manage the business of plant maintenance. Managers need to understand that there are costs associated with collecting, storing, processing, distributing, and using data. However, if it's the right data, and it helps managers run the maintenance business more effectively, it becomes a cost-effective process.

Some of the requirements of good management data are as follows:

- The data must be *necessary* (needed for running the business).
- The data must be *actionable* (in the sense that it indicates situations the manager can influence or control).
- The data must be *accurate* enough to be used for decision-making.
- The data must be *timely* enough to retain its validity.
- The data must be *used*.

Unfortunately, there have been instances where managers have failed to actually use the data, even after it was made available to them. Perhaps this was due to managers not understanding how to use the information or simple feeling they didn't have time to use it. Regardless of the reasons, they were missing out on information that could help them manage more effectively.

Managers are busy and don't have time to read lengthy reports. The most usable management reports are limited to one page (when possible) and presented in a format that makes them easy to read and understand. Above all, they must provide the information the manager needs for decision-making.

There is a broad range of machine history exception reports that can be made available to management for their use. These reports are most useful when there is a focus on the use of graphics and color. Exception reporting should provide a past-to-present record, a comparison of data to goals, and the trend over time. Each report should eliminate the need to retain previous reports.

Popular formats for exception reporting include a minimum of five, or a maximum of ten, machines in the following formats:

- highest cost to maintain
- highest cost to maintain by machine type
- highest downtime hours

- highest downtime hours by machine type
- exceeds costs of same-type machines by X percent
- highest cost to maintain vs machine value
- specific machine/s

In addition to exception reports, there are many other types of standard reports:

- total repair hours
- repair hours by craft
- repair hours by department/craft
- total PM hours
- PM hours by craft
- PM hours by department/craft
- total facility hours
- facility hours by craft
- facility hours by asset type
- total work order hours
- work order hours by department/craft
- total general maintenance hours
- total general maintenance hours by craft
- total maintenance department hours by craft

Machine history reporting isn't limited to just standard reports. Management gets to decide what information they need and in what format they want to see it.

In most manufacturing plants today, managers at all levels, including the first-level foremen, have personal improvement goals and objectives to be achieved. Obviously, the primary objective of the maintenance foreman is to prevent machinery and equipment downtime. That's great! But it's equally important to have improvement goals associated with that objective.

There are specific requirements for objectives and goals. First of all, they must be clearly stated. In addition, each goal must be achievable, measurable, and have a defined a period of time for completion. For example:

Objective: Reduce downtime on no. 2 CNC machine in department 222

Goal: Reduce average monthly downtime by 50 percent by (specific date)

Now that the responsible maintenance foreman knows exactly what's expected, machine history can be used to review various past problems and causes associated with the no. 2 CNC machine and begin corrective actions.

Machine Problem History
#2 CNC Machine Serial No. SV3341G, Dept 222

DATE	PROBLEM	ACTION TAKEN
11-16-17	No Power	Removed solder pool on circuit board
09-14-18	Tool Holder	Replaced tool collet
09-21-18	Overheating	Replaced AC Filter
10-02-18	Z-Axis Servo	Reprogrammed Controller to correct braking problem
10-22-18	Overheating	Cleaned debris from AC Filter
11-14-18	Overheating	Operator Programming Error
01-07-19	Z-Axis Servo	Reprogrammed Controller to correct Braking Problem
04-16-19	Wont Re-boot	Operator Error
07-26-19	Z-Axis Servo	Replaced servo/Reprogrammed Controller
10-20-19	Overheating	Damaged AC Filter Replaced

The results of corrective actions can then be monitored using the machine history system.

A step-by-step process of taking corrective action and then reviewing the results will determine what's working and what isn't. In this way, there will be steady progress toward achieving the goal. In addition, when the foreman's supervisor has questions about progress, machine history will provide the answers.

By this means, over time, everyone from the maintenance skilled trade employees to the plant manager will become aware of the importance of the maintenance machine history and reporting system and its role as a tool for reducing machine downtime. Hopefully, over time, it will be seen even more for its role in reducing the cost of production.

Chapter 17
Education and Training

The development and enhancement of job-related knowledge and skills is vital to plant maintenance due to the continuously changing environment of manufacturing technology. In a typical manufacturing plant, there may be a broad range of training requirements, including the following:

- cognitive and interpretive
- basic technical skills
- advanced technical skills
- machine-specific
- process-specific
- policies, procedures, and practices
- leadership skills

Training presents major challenges for the maintenance manager. How do you get the needed training done with limited time, a limited staff, and a limited budget, and at the same time keep the plant's machinery and equipment up and running? There's also the problem of who needs what training. Everyone isn't at the same level of knowledge and skill. You can't put everyone in a basic technical skills program, because those with advanced knowledge and skills will be frustrated and totally tuned out—a waste of time and money. On the other hand, someone in an advanced program who hasn't yet mastered all of the basics will be out of place—another waste of time and money.

Fortunately, there are commercially available skills assessment tests that will determine the technical knowledge level of each individual. A skills assessment test can answer the question about who needs what training.

There are instances where training is based on seniority: the highest seniority employees get the first opportunity for training. This can result in people being trained who will never use the knowledge and skills they receive, either because their job assignment doesn't require the knowledge or perhaps they are nearing retirement and the value of training will be short-lived.

Fortunately, there are many different education and training resources available, including the following:

- apprenticeship programs
- employee-in-training programs
- trade schools
- college courses
- adult-education classes
- interactive computer programs
- instructional television
- OEM training (onsite and offsite)
- seminars and workshops
- correspondence courses
- lectures and lab sessions
- technical publications

Regardless of the source or form of training, it's absolutely essential for manufacturing plants to keep pace with changing technology, or the inevitable consequences will be higher levels of machine downtime and associated production costs and losses.

Chapter 18
The Evolution of Maintenance Dispatching

The process of assigning work to maintenance employees has been loosely defined as maintenance dispatching. Traditionally, the foreman makes job assignments. At least that has been the method of choice since time immemorial. In fact, who would have thought of it being done any other way?

However, over time, the factory environment has changed, especially in discrete manufacturing, and even more so in large plants. Manufacturing processes have become more complex, requirements have become more demanding, multiple simultaneous machine failures have become commonplace, and the maintenance foreman has found it not only difficult but at times almost impossible to keep on top of rapidly changing events.

There are more than a few things to occupy the foreman's time and attention, especially at the start of the shift, such as reviewing job-status information from the previous shift, answering telephone calls, handling incoming work requests, and making start-of-shift job assignments. Due to these and many other competing shift-start issues, it may take thirty minutes (or longer) before the foreman can get all employees assigned to their first job. In fact, in certain situations, some employees may go an hour or more without a job assignment. The real issue is not so much the wasted indirect labor; it's more about how these delays affect machine downtime, production, and associated costs and losses.

As the shift continues, there are a good many other things the foreman will have to contend with. The following are some of the more common activities:

- attending meeting
- signing requisitions
- following up on purchase orders
- chasing parts
- attending to personnel issues
- dealing with union issues
- handling safety issues
- planning for overtime

- preparing reports
- monitoring priority jobs
- planning work
- making job assignments
- providing jobsite assistance
- answering phone calls and pages
- meeting with vendors
- providing training
- reviewing completed work
- administering discipline

It's easy to understand how hectic the foreman's day can become. This also explains how employees can be idle for extended periods even though they are available for work: the foreman isn't always available to make job assignments.

Some have tried to address this problem by giving employees multiple jobs assignments or having a backlog job rack located where employees who have completed their job assignment would always have another assignment available. However, maintenance requirements are dynamic and ever-changing. The completion of one job assignment doesn't ensure that the next job in the worker's queue is now the most important job, the one that should be assigned first. Priorities in the plant can change quickly, resulting in employees working on lower-priority jobs while higher-priority jobs are waiting to be assigned.

It has long been obvious that the lack of effective and timely communications is a major issue. Early attempts to improve communications in the 1940s came in the form of plant paging systems that used lights or bells. Although this was a step in the right direction, in most cases, it wasn't a practical solution for a factory environment.

The 1950s brought about the installation of strategically located loudspeakers connected to the plant's telephone system. This proved to be significantly better than lights or bells. Even so, there were drawbacks. First, it didn't work well in noisy areas. Second, the person being paged had to find an available telephone somewhere in order to respond, and many times this in itself caused delays. Even though loudspeakers represented some improvement, significant communication problems remained.

A major breakthrough came in the 1960s with the use of handheld two-way radios. Even though early models were both expensive and heavy to carry around, they really did improve communications. The radios were assigned to maintenance foremen and production foremen in key areas. This made it possible for production to contact maintenance almost instantly, resulting in faster job assignments, especially to high-priority jobs such as power outages, machine breakdowns, or other emergencies. For the first time, it could be clearly demonstrated (using response-time data) that management tools like two-way radios could reduce machine downtime. Unfortunately, maintenance machine data was still in the distant future.

General Motors was one of the early pioneers in the use of two-way radios to support maintenance operations in a large manufacturing plant. This process created a single point for all calls for maintenance service. A dispatcher would receive the incoming service requests via telephone and document the necessary job information. This information was then communicated via two-way radio to the maintenance foreman, who would assign the jobs to skilled-trade employees.

Centralized maintenance dispatching using two-way radio communications immediately resulted in faster maintenance response to production requirements. The dispatcher played a key supporting role in making these improvements possible.

In spite of the many benefits of centralized maintenance dispatching, many of these early dispatching systems ultimately failed. In most cases, it was due to the need to reduce headcount during times of austerity. Unfortunately, it was likely assumed that removing the dispatcher to achieve headcount reduction wouldn't be a major issue, because the productivity improvements would continue through the use of two-way radio communications.

Even as many of these early systems were being reduced or eliminated, the Delco Radio division of GM, which later became the Delco Electronics division, was taking some innovative steps that would make centralized maintenance dispatching a far more effective process. This dispatching system incorporated what is believed to be a first in a manufacturing plant: direct radio dispatching of maintenance skilled-trade employees to support production operations. This created a fast and effective radio communications link between production, the dispatcher, the foremen, and their maintenance skilled-trade employees.

Delco also piloted the use of computer data cards, which were batch-processed to produce machine history and other maintenance reports. This dispatching system was planned for implementation in two phases. The first phase included installing a dispatch office in one of four local plant locations.

This also became a pilot for proving the effectiveness of (non-supervisor) dispatchers making job assignments to maintenance employees. Although initially, during the start-up, the first dispatcher was a maintenance foreman, as the process expanded, clerical employees were trained as dispatchers. This proved to be successful, and future dispatchers were all from a nonmaintenance background.

In order to avoid conflicts with the local union, it was made clear from the beginning that the hourly skilled-trade employees worked for their foreman, not the dispatcher. The dispatcher was simply assisting the foreman by communicating job assignments. The foreman could direct the dispatcher in making or changing job assignments at any time. Shown below is a picture of the first dispatching station.

The benefits of this new system soon became obvious to its production customers, who strongly supported it to upper management. This led to a data-processing project to develop the machine-history reporting system. Twelve months after the initial start-up of the dispatching center, the reporting system was ready. Each morning, completed data cards for the previous day were sent to data processing for key-punching. Monthly reports were generated containing the following data:

- machine downtime by plant, department, and machine
- repair hours by plant, department, and machine
- repair hours by craft, plant, department, and machine

In 1977, armed with the ongoing success of the Centralized Maintenance Dispatching System, Delco began planning for a far more advanced system, one that would provide a level of real-time dispatching that could not be achieved using a card system to manage job assignments. The new system would incorporate a centrally located dispatching center that could support the requirements of its three local plants.

The new system, when it began operation in 1979, was probably the most advanced of its kind in existence dedicated solely to maintenance dispatching; it integrated the use of radios, telephones, and computers into a truly real-time dispatching process. The dispatching console with data entry and display terminals was connected to an on-site host minicomputer, which was interconnected to an off-site mainframe computer.

This configuration enabled the dispatcher to receive a request for service via telephone, enter the job information into the data entry terminal, scan the monitors to identify the appropriate employee for the assignment, and then assign the employee via two-way radio. All of this could be done in less than one minute. When employees completed a job assignment, they notified the dispatcher via radio and provided coded job-completion data. They could then be reassigned to another job, all within thirty seconds or less.

This made possible a level of maintenance resource management that could not otherwise have been achieved in the dynamic environment of a large manufacturing plant, where there were more than five thousand production machines at three local plant locations. This included standard machine tools, sophisticated chip-making equipment, computerized electronic testing, even robots. Obviously, multiple and simultaneous machine downtime was commonplace. The dispatching center is shown below:

This was the kind of dynamic ever-changing environment that real-time dispatching was designed for, ensuring the continuous assignment of maintenance resources to the most important jobs first. Doing this requires reassigning maintenance employees from lower-priority in-progress jobs to more-important higher-priority jobs, especially during peak demand when it's critical for everyone to be continuously working on what is most important at the time.

In addition to the dispatching center, there were remote terminals located in specific high-tech manufacturing areas where new job requests could be entered by production. There were also restricted high-tech areas where the maintenance employees could use the terminals to assign themselves to jobs within that area. Maintenance foremen could use any terminal in the plant to periodically review current job assignments, job backlog, priorities, etc. When necessary, a foreman would provide guidance to the dispatcher in making specific job assignments or in determining the order in which jobs would be assigned. Due to the close working relationship between the foremen and the dispatchers, no problems ever developed.

Nevertheless, few companies would have made the kind of investment in their maintenance department to fund a real-time dispatching system like the one described here, even if it were economically justified. But be glad you waited!

Continuing changes in the world of computer technology have come to the rescue again. What would have cost a lot of time and money just a few years ago can now be had for pennies by comparison.

Some of the personal computers of today have greater computing power and speed than the large mainframe computers of the 1980s. These desktop machines costing $2,000 or less are capable of supporting a real-time maintenance dispatching system. The required software can be self-developed using a low-cost commercial database manager.

Shown below is a dispatching center installed in a medium-size manufacturing plant that began operation in 1999.

Maintenance employees were dispatched here via two-way radio. The design and configuration of the dispatcher workstation focused on low cost, simplicity, and utility. Preventive maintenance schedules were generated at the same workstation. The estimated ROI was four months, with cost savings and cost avoidance to be achieved through increased worker productivity and reduced machine downtime, including associated production costs and losses.

A similar dispatching system was implemented at the GM electromotive plant in 1993. This plant, which manufactured diesel locomotive components, also modified and rebuilt used locomotives. Centralized dispatching was piloted first in one critical manufacturing area where machine downtime was historically high. During the first three months of operation, machine downtime was reduced 36.9 percent compared to the previous four months for which downtime data was available. Similar (or even better) results can be expected when centralized maintenance dispatching is properly designed and implemented within the larger context of planned failure response.

Centralized maintenance dispatching can be designed and configured to meet the needs of almost any plant, large or small. Thanks to the use of mostly off-the-shelf radio and computer equipment, cost is no longer a significant issue. The real issue of significance is the cost benefits, some of which are listed below:

- improved communications
- elimination of lost or misplaced work requests
- faster responses to emergencies and high priority jobs
- continuous assignment of workers to the most important jobs
- increased worker efficiency and tool-in-hand productivity
- improved management of priorities, workload, and resources
- reduced downtime and associated production costs and losses
- improved foreman efficiency and oversight effectiveness
- data for preventive maintenance activities
- actionable machine history data
- resource planning data
- data for budget planning

Remember, it takes more than technology to make centralized maintenance dispatching or any other maintenance strategy successful. It also requires appropriate changes in organizational structure, systems, processes, policies, procedures, and practices, plus a good bit of training. In addition, if people are not made accountable for results, all of the effort will be in vain.

Chapter 19
Maintenance Hourly Employee Survey

It's important for the maintenance department to understand how hourly employees feel about their work environment. Feedback is important. If management is doing a good job, that's great. If not, what are the issues? What if there are problems that management isn't aware of? What if there is just a perception of problems that really don't exist? Remember, if someone thinks there's a problem, that's a problem!

While management may be talking to hourly employees every day, the feedback they are getting may not always be all that it should be. Sometimes, people just don't want to be seen as complainers. Others might not complain because they like their manager and don't want to say anything that could affect a good working relationship.

Surveys can be powerful tools in bridging the gap between what people don't say and what they really need to be saying. Surveys give people the opportunity to speak anonymously. Surveys can give the maintenance department an opportunity to correct problems they aren't aware of, or bring understanding where there may be misunderstandings. Remember the results of the famous study conducted in 1927 at the Hawthorne Works of Western Electric in Chicago: "Show some attention to the worker and the worker will take care of the job!"

The following survey (or something similar designed in-house) can be used to solicit hourly employee input:

Maintenance Hourly Employee Survey

Additional Comments can be written on the back of this page

Mark an X in the box of your choice	Agree	Neutral	Disagree
I enjoy working here because of the work I'm engaged in.			
I feel like I am appreciated as a worker and as an individual.			
I work in an environment of trust and respect.			
Management treats people equally and fairly.			
Management never criticizes an employee in the presence of others.			
I feel like I have equal opportunity with others for advancement.			
Management solicits my ideas for making my work easier and safer.			
If I have questions management takes the time to answer them.			
I feel free to disagree with things that may adversely affect me.			
Management always gives me the opportunity to talk in private.			
Management provides the training I need to get the job done right.			
Management provides the special tools needed to get the job done.			

Maintenance Shop Survey

There are numerous contributors to worker productivity and quality workmanship. First of all, an employee must have the technical knowledge and skills to perform the work. When an employee is performing service work out in the production areas of the plant, technical knowledge and skills are complemented by the needed spare parts, materials, supplies, portable tools, and equipment. SPM focuses on management controls that can make this complex process both effective and efficient.

The maintenance shop is a much different work environment than performing service support to production. However, the maintenance shop is a key resource supporting this service work. In general, shop work requires a different set of technical knowledge and skills. However, in shop work, productivity and quality workmanship are greatly dependent upon the machinery, tools, and equipment made available to these workers. The maintenance shop includes tool, fixture, and die repair.

While no specific timeline exists, it's probably a good idea to perform a shop survey if the shop has been in operation five years or longer, and every five years thereafter. The objective of a shop survey is to determine the operational condition, versatility, and availability of each capital asset in the shop. These are critical asset performance factors. The survey is a tool that can help identify areas where management intervention may be needed to ensure these physical assets are continuing to be positive contributors to worker productivity, quality workmanship, and (their ultimate influence) the cost of production.

The survey itself is best conducted by a neutral party outside of the maintenance department, working together in close cooperation with the maintenance manager. An engineer with manufacturing experience would be an excellent choice. Input to the survey is provided by the maintenance employees who regularly use the asset or assets being surveyed. The maintenance manager will select participants based on ability to provide objective input.

A second benefit of the shop survey is the visible evidence of management's interest in employees having the machinery, tools, and equipment they need to get the job done right the first time, on time, and at the lowest cost. This is in the best interest of both the worker and the company. Giving employees the opportunity to provide input to management is always the right thing to do.

Care should be taken to not create unreasonable expectations that problems found can be quickly corrected. The message must be clear that the survey is exploratory, and if problems are found, they will be reviewed by management to determine what corrective actions are needed and if the cost can be justified. In addition, the employees involved will receive feedback in a reasonable timeframe. There should be ready answers to the questions that may arise about survey results and actions to be taken, if any.

The survey is simple and can be completed in a relatively short period of time. The number of individual assets will vary by craft group. For example, there may be few physical assets in the pipe shop, but there will probably be many more in the machine shop. The survey is conducted as follows:

Step 1: Obtain from asset accounting a listing of all capital assets assigned to the maintenance department. If such a listing is not readily available, the maintenance manager can assign someone in each craft shop to make such a listing, or the asset listing can be obtained by the participants during the survey. The required information includes asset number, asset name, manufacturer, and model number, when available.

Step 2: Two forms are needed for use in the survey. The first is used by the employees participating in the survey; it's shown below and can be printed on four-by-six card stock.

PHYSICAL ASSET SURVEY QUESTIONS

Respond to each question using the numbers 0, 1, 2, 3, or 4

1. **CONDITION**
   How do you rate the current operating condition of this asset?

   4) Excellent   3) Good   2) Fair   1) Poor   0) Inoperable

2. **VERSATILITY**
   How well do the functions and features of this asset meet my needs?

Each response has a numeric value of 4, 3, 2, 1, or 0. The survey values are used to call management's attention to areas where quality workmanship and productivity may be adversely affected. A second form, the survey worksheet, is used to document the survey responses.

For individual assets, a 2 or lower value on any of the three questions raises a red flag. Obviously, any value of 1 or lower is reason for management to look into the problem as soon as possible. The survey will not only establish values for each individual asset, the values will be summarized for each craft shop and then for all maintenance shops. The survey process and objectives are discussed with the participants to answer any survey-related questions.

Whoever is conducting the survey will discuss these forms and how they are used with everyone involved prior to the survey. Discussing the use of the forms is a good way to solicit questions that need to be answered before proceeding. This step is essential for ensuring good logical survey responses.

The second form, the shop survey summary worksheet, appears on the following page.

#	ASSET #	SHOP SURVEY DATA ENTRY FORM (4 • 3 • 2 • 1 • 0)			
		ASSET DESCRIPTION	Q1	Q2	Q3
1					
2					
3					
4					
5					
6					
7					
8					
9					
10					
11					
12					
13					
14					
15					
16					
17					
18					
19					
20					
21					
22					
23					
24					
25					
26					

Maintenance Customer Satisfaction Survey

It's important for the maintenance department to understand how production customers feel about the service they are getting. If maintenance is doing a good job, that's nice to know. It's something to be shared with others.

What if there are problems the maintenance department isn't being made aware of—or what if there is just a perception of problems that really don't exist? Remember, as mentioned above, if someone thinks there's a problem, that's a problem.

While it's important to talk to production customers face-to-face, the feedback may not always be what it should be. Sometimes people just don't want to be seen as complainers. Others might not complain because they like the maintenance people they work with and don't want to offend them. Still other may not complain because they fear their service might suffer.

Surveys can be powerful tools in bridging the gap between what people don't say and what they need to be saying. Surveys give people the opportunity to speak anonymously. They can give the maintenance department an opportunity to correct problems they aren't aware of and bring understanding where there may be misunderstandings.

Surveys can be used to improve the effectiveness of the services provided to production customers. The following survey, which has been used by maintenance departments in the past, is included as a thought starter.

Maintenance Customer Satisfaction Survey

Mark an X in the box of your choice	Agree	Neutral	Disagree
There is a fast and convenient process for requesting maintenance services.			
Service requests are rarely misplaced, forgotten, or lost.			
When a machine is down maintenance responds quickly.			
Repeat calls to fix the same problems are rare.			
Maintenance is careful not to damage production parts near the work-site.			
When work is complete the jobsite is left clean, orderly & ready for production.			
Production is kept informed about the status of ongoing repairs.			
Once a job starts it's rarely delayed due to the lack of parts, tools, or manpower.			

Reverse Appraisal

One of the most important things in any organization is good communications. All too often, it's the lack thereof that can have a negative influence on the effectiveness of the organization. On the other hand, effective communication can be a powerful tool for creating and maintaining organizational effectiveness and a quality work environment.

One highly effective communication tool is called the *reverse appraisal*. Organizations have been successfully using some form of this process to improve communications for years; it's not something new. In the past, reverse appraisals have been used almost exclusively for managers, but there is reason to believe that it could be just as successful for first-line supervisors and their hourly employees.

The reverse appraisal is a facilitated communication exercise between a superior and subordinates. In a reverse appraisal, the roles are reversed, and the subordinates, with the assistance of a trained facilitator, conduct an appraisal of their superior. A 4:1 or greater ratio of subordinate-to-superior is recommended for this exercise. The process uses a questionnaire in which the participants are asked just six questions.

It's important that the reverse appraisal be introduced to the organization in a positive way that dispels possible apprehensions about being appraised by your employees. One way to accomplish this is to start the reverse-appraisal process with a senior manager who then can introduce it to the larger organization by sharing personal experience with the process. Afterward, the process can work itself downward through the organization.

It should be emphasized that the results of a reverse appraisal are totally confidential; they are not reported, disseminated, discussed, or shared with anyone outside of the group involved, unless the manager voluntarily chooses to do so. Managers should never feel apprehensive about participating in a reverse appraisal; the objective is to improve communication and working relationships among everyone within the work group.

The following are typical manager comments after participating in a reverse appraisal:

"I never realized there was such a lack of communication among us."

"I'm really glad we did this."

"I wish I had known before what I know now."

"This is going to be a real help to me; to all of us."

"I want to do it again in six months to see where we are then."

Hints for the Manager before the Reverse Appraisal

- Relax! This helps your employees to relax and feel free to participate.

- Solicit and expect open, candid, forthright feedback.

- In the event of a negative comment, avoid becoming defensive. Negative feedback can be really important, and defensiveness can get in the way of that information.

- Remember, any problem (whether it's real or not) is a real problem as long as there is a perception that a problem exists.

- Seemingly insignificant comments can be important to those who make them. Treat every comment with a careful response.

- While the focus of the reverse appraisal is on just six questions, the discussions can and probably will migrate to unrelated issues. This should be controlled but not discouraged, because discussing those issues could be important to better communications.

- If there is disagreement with a comment, take time to fully explain why you disagree. It's important to respond as clearly and completely as possible to all of the employee comments. If additional time is needed, extend the meeting. If that's not possible, schedule another meeting.

- If information isn't available to properly respond to a comment or question, a commitment should be made to respond at a later time.

- If the reverse appraisal results in action items for yourself or others, schedule a follow-up meeting to review progress or results.

The Reverse Appraisal Process

Step One

The facilitator and the manager to be appraised meet to review the reverse appraisal process and the associated six-part questionnaire. The purpose is to do the following:

- discuss why the appraisal is being conducted
- review the objectives and benefits of conducting a reverse appraisal
- explain the manager's role in making the appraisal successful
- answer the manager's questions and concerns
- try to avoid unnecessary apprehension on the part of the manager.

A copy of the reverse appraisal questionnaire is shown below. The blank in each question is space for the manager's name.

Reverse Appraisal Questionnaire

1. What could _____ do to manage more effectively?
2. What could I do to help _____ manage more effectively?
3. What could _____ do to improve the effectiveness of our work-group?
4. What could I do to improve the effectiveness of our work-group?
5. What are the things that _____ does that I really don't like and I wish he/she would stop doing them?
6. What are the things that _____ does that I really like and I hope he/she continues to do them?

Step Two

A meeting is scheduled with everyone who directly reports to the manager. The manager opens the meeting by introducing the facilitator and informing the group that they will be participating in a reverse appraisal by completing a questionnaire, which the facilitator will further explain. After encouraging the group to be open, frank, and forthright in all of their responses to the questionnaire, the manager leaves the meeting.

Step Three

The facilitator will explain why a reverse appraisal is being conducted before passing out copies of the questionnaire. When the participants have had time to review the questionnaire and ask their own questions, the following points will be explained in detail:

- Only the facilitator sees the completed questionnaires. The responses are copied and reformatted into a single document, and then the original questionnaires are destroyed.
- Professionalism in responses is expected. Any demeaning remarks directed at the manager or a coworker will be deleted. Respond as professionals!
- All responses will remain anonymous. Anything that could potentially identify the submitter will be deleted.
- Multiple similar responses will be rewritten as a single response that was submitted by a number of participants.
- The schedule for a group meeting in which the manager will respond to the completed questionnaires is set.

Step Four

The completed questionnaires are assembled into a single reformatted document, and the original copies are destroyed.

Step Five

The facilitator meets with the manager at least one day prior to the scheduled reverse appraisal response meeting. This opportunity to review the reformatted responses is essential to allow the manager time to review the responses and prepare information that might be needed for the meeting. The manager is cautioned not to discuss with employees anything related to their responses prior to the scheduled meeting.

Step Six

The facilitator, manager, and employees meet for the reverse appraisal review. The facilitator begins this meeting with a brief review of the RA process to this point, answers relevant questions, and discusses any rules or guidelines that may be established for the meeting.

An overhead projector or other media device is used to display the six-part questionnaire. The first question and first response to that question are displayed. At this point, the manager responds to that question.

The manager can ask for clarification to better understand the question, the response, or both. At the conclusion of the manager's remarks, the facilitator asks the group if the manager has satisfactorily addressed the subject. If not, a two-way dialog takes place between the manager and employees until the matter is satisfactorily addressed.

The second response to the first question is now displayed for the manager's response. This process is repeated until all of the responses to the first question have been addressed. At that point, the second question and the first response to that question are displayed, and the process continues until all of the questions and responses have been addressed.

Prior to concluding the meeting, the employee group will complete a short reverse appraisal evaluation form. Afterward, the facilitator will discuss the reverse appraisal meeting with the manager to get comments about the process itself and any changes or improvements they might suggest. The manager is provided with all of the associated paperwork for personal use.

A copy of the reverse appraisal evaluation form is shown below.

Reverse Appraisal Evaluation

(Please circle your answers)

1. Did your manager satisfactorily address your questionnaire responses?

   YES   NO   If no, please explain. _____

2. Did you feel free to be open, honest, and frank in your responses?

   YES   NO   If no, please explain. _____

3. Did you feel free to be open, honest, and frank in the meeting discussions?

   YES   NO   If no, please explain. _____

4. Did communication take place in the meeting that may not have taken place otherwise?

   YES   NO   If no, please explain. _____

5. How would you rate your overall reverse-appraisal experience?

   Excellent   Above average   Average   Below average   Poor

Comments (Use the back of this form if needed)

Chapter 20
System Concepts 101

The word *system* seems to have originated in the realm of academia and the sciences; the word wasn't commonly used outside of that arena for hundreds of years. However, today we have become increasingly aware of systems in our everyday lives. There are monetary systems, educational systems, transportation systems, judicial systems, manufacturing systems, management systems, and the list goes on and on. Systems are being used throughout business and industry to manage their operations.

This is also true in plant maintenance. Since plant maintenance has a role in the success or failure of the larger business enterprise, it's important to understand the role of systems, because dysfunctional plant systems are one of the leading causes of excessive maintenance costs and maintenance-related production costs and losses.

What Is a System?

What are the attributes and characteristics of a system, and what makes a system functional or dysfunctional? The answers to these questions require, at the least, a basic understanding of certain system concepts. Georg Wilhelm Friedrich Hegel (1770–1831) is considered to be the father of modern systems thinking. He developed what is known as *general systems theory*. The postulates (or rules) of Hegel's theory are as follows:

- The whole is more than the sum of the parts.
- The whole determines the nature of the parts.
- The parts cannot be understood if considered in isolation from the whole.
- The parts are dynamically interrelated and interdependent.

Theory may not seem very important to those who are struggling with the day to day problems associated with manufacturing, but Mr. Hegel's theory provides an insight into how dysfunctional plant systems can drain the economic vitality of a manufacturing plant. The techniques used here to help busy managers better understand system concepts may seem overly simplistic for such a complex issue. However, there are no apologies for doing this if 1) it promotes a greater understanding about system concepts; 2) it helps people recognize the consequences of dysfunctional systems; and 3) it leads to developing more effective systems that become enablers to the goals and objectives of the business enterprise.

Some systems are unbelievably complex, while others are quite simple. For example, a jigsaw puzzle is a very simplistic system, one that most people are familiar with. There is a popular 1,500-piece jigsaw puzzle of a colorful woodland scene. Consider how this puzzle conforms to the rules of general systems theory:

- **The whole is more than the sum of the parts.** The puzzle pieces themselves are nothing more than a group of colored, irregularly shaped parts ... until they are assembled together with the other parts. When the puzzle is fully assembled, something more than 1,500 individual parts exists—a beautiful woodland scene with trees, flowers, rocks, and a small stream with its reflecting lights and colors has been added.

- **The whole determines the nature of the parts.** It's only after the puzzle has been fully assembled that it is clearly demonstrated how the complete puzzle has determined the role of the individual parts.

- **The parts cannot be understood if considered in isolation from the whole.** The individual puzzle parts are nothing but irregularly shaped pieces of paperboard with colors on one side. It's only after they are joined together with the other parts that the role of each part can be understood.

- **The parts are dynamically interrelated and interdependent.** The individual size, shape, and color form of each puzzle part determines its relationship to, and dependency upon, the other puzzle parts as they are joined together to form the whole puzzle.

System Concepts

Now consider how Mr. Hegel's theory applies to a much more complex system, such as an automobile. An automobile is comprised of thousands of individual pieces and parts that must not only fit together but work together. It is both in fitting together and working together that things most often go wrong. Adding certain system concepts to the general systems theory may help in understanding how system characteristics determine how well or how poorly the individual parts fit together and work together.

Systems Concept #1: *If the system as a whole is operating at its best, the individual parts of the system probably are not.*

If this concept appears to be a contradiction, it really isn't. This can be demonstrated by using a little imagination to create some what-if scenarios. What if the (imaginary) XYZ Automotive Testing Laboratory has been contracted to perform two different automobile performance tests? These tests will be performed on standard front-wheel drive, full-size, four-door sedans, excluding premium and high-performance cars. All brands sold in the United States will be tested.

The first test is to determine the best car. Each car is scored using a standard group of performance indices, such as:

- fit and finish
- ride and handling
- acceleration and braking
- visibility
- noise
- ease of use of controls
- safety features
- temperature control
- seat comfort
- rear-seat legroom
- cargo space
- fuel consumption,
- owner satisfaction

- predicted reliability
- cost

Based on the results of these individual tests, each car receives a total score. The highest total score will determine the best car.

Once this phase of testing has been concluded and the best car has been identified, a question is asked: does the selected best car also have the best individual parts and components that could have been found among all of the cars tested? What are the possibilities of that ever happening?

Systems Concept #2: *If the individual parts of the system are all the best that can be obtained, the system as a whole probably is not.*

The second-phase of testing by the XYZ Automotive Testing Laboratory will take quite a bit longer, because they will be testing all of the cars again, except this time, they are searching for all of the best individual parts and components needed to build an automobile. For example, they are searching for the best body/cabin, engine, transmission, chassis and suspension, steering mechanism, starter, generator, water pump, etc.

After a period of time, the monumental task of identifying all of the individual best parts needed to build an automobile is completed. Unfortunately, this wonderful collection of best parts couldn't possibly be used to build an automobile; because they were never designed to fit together or work-together.

System Concept #3: *The performance of the whole system is not determined simply by assembling a group of the best parts; rather, it's a consequence of how well all of the parts fit together and how well they work-together.*

There are, in virtually every manufacturing plant, a lot of exceptionally good pieces and parts, such as individuals, work groups, and departments. Unfortunately, due to the unique systems designed specifically for each individual work group and department, they don't always fit together and work together as effectively as they should with all of the other plant systems they interact with. This occurs when systems are designed without consideration for all of the different fitting-together and working-together requirements in a manufacturing plant.

Systems Concept #4: *Events occur within systems as a result of many forces working together in complex relationships with one another.*

In bygone days, it wasn't uncommon for boys to carry a few acorns in their pockets. Acorns made good flipper loads. Also, they were super for throwing at anything that suited their fancy. There seemed to be a natural affinity between boys and acorns. However, acorns have a more important purpose in nature: they are necessary for the propagation of oak trees.

There are cause-and-effect relationships taking place in the creation of an oak tree. In the case of the acorn, while it's necessary, it's not sufficient in itself to create an oak tree. Other things are also required, things which are both necessary and sufficient to produce an oak tree. These include a seed bed, moisture, heat, sunlight, and time. All of these entities work together in complex relationships with one another in order for the acorn to become an oak tree.

Similarly, there are system issues that affect every organization. Many forces work together in complex relationships with one another to influence the effectiveness (or ineffectiveness) of the organization. Unfortunately, when these complex relationships become dysfunctional, they lead to organizational constraints that manifest themselves in excessive costs. These situations can be corrected, but it takes a greater level of systems thinking in the design, development, and implementation of systems and their internal management controls.

There seems to be a natural opposite tendency toward linear-thinking, which views most problems as having single causes and single solutions. Systems thinking doesn't seem to come naturally; it must be nurtured.

Systems Concept #5: *Systems-thinking assumes that all problems are embedded within a larger problem situation that has multiple causes, and any actions taken to correct the problem can have consequences both intended and unintended.*

The lack of systems thinking is easily demonstrated using a real-life problem that is all too common in the maintenance department: the procurement of machine repair parts. Purchasing is the key procurement function in a manufacturing plant. There are specific policies, procedures, and guidelines that define exactly how the business of purchasing is conducted. Purchasing buyers process purchase requests; select vendors based on various criteria including pricing, terms, and delivery; execute purchase orders; provide follow-up, etc.

When a machine replacement part is requested, the only available source may be the original equipment manufacturer (OEM). When there are multiple sources for the same type of item, the buyer may choose the source based on criteria like quality, price, and delivery. In this example, a purchase request was made by the maintenance department for a relatively expensive sealed and shielded tapered roller bearing. The OEM part number was included in the request. The purchasing department buyer found that there were multiple sources for the item requested. The buyer then selected a vendor and issued a purchase order based on price and promised delivery.

When the bearing was delivered to the maintenance department two days later, it was quickly discovered that it wasn't the OEM part that had been requested. The maintenance department immediately contacted the purchasing department about the problem. The buyer defended his decision by declaring that he had saved the company more than fifty dollars by purchasing the generic version of the bearing. It was only then that the buyer learned that generic bearings had been tried before and had failed prematurely. The bearing had to be the OEM part number originally requested.

This fiasco resulted in two additional days of costly downtime, plus the expensive bearing that wouldn't be used. All of this could have been avoided by have a simple notation on the purchase request: *Do not substitute*. When the purchasing department designed its system, it failed to get input from the maintenance department—a major customer of the purchasing department.

In this example, maintenance and purchasing were actually both customers and suppliers to each other. The maintenance department was a supplier of purchase requests to purchasing and a customer for items purchased. Conversely, the purchasing department was a customer for purchase requests and a supplier of items purchased.

Unfortunately, these kinds of system failures really do occur, and all too frequently. The majority of these kinds of failures can be prevented with careful consideration for all of the potential customer-supplier relationships that have to fit together and work together in order to be successful.

Systems Concept #6: *Changes to the existing system are filled with opportunities for unexpected and unwanted consequences.*

Even though there is risk involved in any change, if and when the change process is driven by systems thinking, adequate planning, and careful execution, then positive change can take place. When there is adequate consideration of the many interrelated and interdependent activities that take place between autonomous groups, the larger organization of which they are a part will also become more synchronous and more successful.

Chapter 21
The Challenges of Positive Change

Why do we hear so much about change today? Why is change so important to us? Is it because everything around us seems to be changing? Are things we may have taken for granted in the past—institutions, values, relationships, things that we considered enduring—giving way to change?

In addition to change itself, the rate of change seems to be accelerating. Sometimes the reasons for change seem subtle or ambiguous. At other times, the reasons are all too clear: we must change to survive.

One of the most challenging aspects of management is the management of change. What makes it so challenging is understanding that change can impact the very fabric of the organization and its members, goals, and objectives. Its very future may seem threatened. Also, there may be unintended consequences to change. The final results of change, for better or for worse, depend on just how well or how poorly the process of change is managed.

When change is going to affect people in the organization, how are they going to react to the change? There are several things that influence how people react to change:

- the existing organizational climate
- the perception of the need for change
- the perception of how change will affect individuals
- how the change process is managed

People's attitude toward change can affect their productivity due to the stress commonly associated with change. However, most of the trauma and stress could be avoided if the same basic principles used to manage other aspects of the business were applied to the process of change.

Any good sports coach will tell you that consistent winning depends most of all on effectively executing the basics, the fundamentals of the game. The message is simple: stick to the basics. What are the basics of change? Dealing with common perceptions about change might be a good place to start.

People resist change, right? *Wrong.* People want change. They are asking for change. Ask the skilled maintenance employees if there are any changes they would like to see; be prepared for more than enough.

Did people resist the change from fountain pens to ballpoint pens? Did they resist the change from slide rules to calculators? Why is it that people buy a new car when their current car is still fully capable of meeting their transportation needs? It's not change itself that people resist; it's *being* changed. People resist change that's imposed on them, change they don't understand, change they don't see the need for, and change they fear will adversely affect them.

It's managements role to make strategic business decisions that are in the best interests of the business enterprise. That may take the form of changes to organizational structure, systems, processes, policies, procedures, and practices. These changes inevitably affect people in the organization. It's important that everyone affected by these changes becomes aware of the change and understands and accepts the need for it.

Most people can understand and accept the need for change when the reasons are clearly explained. However, there are some important dos and don'ts. *Do* stress the positive aspects, the benefits of change. *Don't* hide any of the negative aspects of change. Be honest and up front; tell it as it is. It must pass the test of logic and reason.

The following was written long ago:

> There is nothing more difficult to take in hand, more perilous to conduct, or more uncertain in its success, than to take the lead in the introduction of a new order of things, because the innovator has for enemies all those who have done well under the old conditions and lukewarm defenders in those who may do well under the new. (Machiavelli, *The Prince*)

Regardless of Machiavelli, if managers do their job in clearly explaining the need for change, it will be received by most people with understanding and acceptance.

However, management needs more than just awareness, understanding, and acceptance. There must also be a commitment to change, and successfully getting to a commitment can take the form of a six-step process:

1. Awareness
2. Understanding
3. Acceptance
4. Participation
5. Ownership
6. Commitment

Even though people have already become aware of the change and understand and accept the need, it's not uncommon to hear people complain that management doesn't involve them until after most or all of the decisions involved in the change have already been made.

By allowing people to at least influence some of the decisions, especially about things that directly or indirectly involve them, management can develop a sense of ownership in bringing about the change. Inevitably, that sense of ownership and commitment helps to make the change successful. This process isn't rocket science; it's just common sense.

Chapter 22
Common-Sense Indices of Business Performance

Every business enterprise needs a way to gauge how well or how poorly it is performing. The standard method of determining the health of a business has historically been to measure bottom-line financial performance. This seems to make perfectly good sense if a company wants to stay in business. This isn't to say that other factors aren't important to the success of a business, but the bottom line is still all-important.

In the 1980s, something rather amazing happened: management consultants discovered the term *world-class*, and in the process ushered in a new era of consulting opportunities. *World-class* became the latest in a long list of buzzwords. This was followed by numerous new management books touting the idea of becoming world-class.

The concept of world-class introduced new dimensions for measuring business performance. Consultants formalized the concept of world-class by obtaining data from companies considered to be highly successful. Based on the initial data, the number of companies was reduced to just those where a comprehensive analysis of the company and its operations could be performed.

Ultimately, certain companies were recognized as leaders in their line of business because of continuing long-term growth, increased value to their customers, and outstanding financial returns to their investors. These companies were identified as being world-class.

This endeavor focused on identifying the unique attributes and characteristics of these companies and their operations—those factors that contributed most to their success. These attributes and characteristics were transformed into indices that were packaged into products and services that would enable other companies to see how they measured up to world-class. In addition, there was the usual consulting support to help other companies focus on the things that would help them to become world-class.

These endeavors, plus the expanded media coverage, was all that was needed. It wasn't long before almost every company wanted to become world-class. That's exactly what the consulting industry was looking for. Not that there's anything wrong with consulting support, especially if you are dealing in areas where you have no expertise. What you don't need and don't want is someone coming in with a set of canned solutions without first identifying the problems that exist within your company.

If you opt for consulting support, your best bet is to have more than one consulting company look at your situation (at their expense) and bid for the business of providing a solution. That's a common-sense approach which may not be to the consultants' liking. On the other hand, it's your money, and you decide where and how to spend it.

The bottom line is (and this is important), how will the improvement be measured? That's something that must be established and agreed upon up front. All kinds of metrics have been established to measure performance. However, if the metrics involved don't establish some measurable cost relationship between maintenance performance and the cost of production, then maintenance will never be seen as anything more than a cost, rather than a contributor to the cost objectives of the business enterprise.

The IMPACT Business Case

The maintenance department can boast about having strong technical skills. On the other hand, maintenance departments may have limited business skills; it definitely hasn't proved to be one of their stronger suites. For example, the maintenance department has often been weak in putting together a good business case that justifies (appropriated) spending outside of the maintenance budget.

Many times, first-step funding approval has been based more on trust and confidence in the requestor's judgement than a good business case, leaving it to others to develop the required business case. This is unfortunate, because there are ways the maintenance department can put together a good business case in a language that financial people can understand and appreciate.

In fact, one large US manufacturing plant, with the help of an outside maintenance consultant, did just that: they used a business-case approach that I have dubbed the IMPACT business case. The name is appropriate because it so clearly demonstrates the impact that maintenance costs have on the larger business enterprise. It also demonstrates the magnitude of potential savings.

The maintenance department in this plant, with the assistance of a facilitator, identified a set of cost contributors that could be managed to generate major cost avoidance and cost savings. Using published industry data, they estimated the total cost of maintenance for their plant and developed a business plan with estimated savings.

The strategy for reducing maintenance costs was focused on reducing or (where possible) eliminating on-the-job work constraints, roadblocks, barriers, and delays that were hampering the ability of the maintenance department to effectively support production operations. These constraints were increasing both repair time and machine downtime; this, in turn, increased the cost of production.

The first step was to identify the most common on-the-job work constraints faced on a daily basis. A one-day workshop was scheduled for this purpose. Although published industry estimates were available, the objective here was to develop realistic in-house estimates of the maintenance labor being wasted due to these constraints.

A special project team was formed for the workshop. The team consisted of both salaried and hourly maintenance employees. Included were a maintenance supervisor, three first-level maintenance foremen, eight skilled-trade employees who represented the various craft groups, plus the union skilled-trade committeeman. All of the participants had worked for the company five years or more, and everyone had volunteered to participate. The workshop was held in a conference room away from the noise and distractions of the plant operations. The participants were made aware in advance that the data they were developing would be presented to the plant manager and his staff, and possibly to corporate management.

The first workshop activity involved identifying the work constraints that were most prevalent and causing the most problems. A follow-up step would estimate how much actual maintenance-department labor was being wasted due to these constraints. The participants were cautioned again and again about taking time to provide thoughtful, carefully considered responses, making sure their estimates were as realistic. This repeated cautioning about carefully considered estimates was felt to be key to fostering a sense of responsibility for the task they faced. This also increased the probability that their estimates would pass the tests of logic and reason.

Ultimately, the following major on-the-job work constraints were identified (listed here in alphabetical order):

- communication problems
- excessive travel to and from the jobsite
- inadequate and inaccurate job information
- lack of technical training
- spare parts and material shortages and outages
- waiting at the spare parts crib
- waiting on additional craft assistance
- waiting on engineering assistance
- waiting on operator assistance
- waiting on requisition approval
- waiting on safety-related permits
- waiting on supervisory assistance

Considerable time was taken to discuss each of the identified constraints. Then, each participant submitted an estimate of the average total hours lost by each maintenance employee during an eight-hour shift due to all of these work constraints combined. The constraints were then weighted to establish priorities for corrective action.

The results of the project team estimates were as follows:

- low estimate—1.5 hours (discarded)
- high estimate—4.0 hours (discarded)
- group average—2.8 hours (35 percent of an eight-hour shift)

If the estimate of 2.8 hours a day lost by each employee seems high, then consider the following published industry estimate shared by Terry Wireman in his book *World Class Maintenance Management* (Industrial Press Inc., 1990): "Less than

4hr/day (out of a possible 8) are spent by maintenance craftsmen performing hands-on work activities."

The second workshop activity identified some of the most common maintenance-related costs and losses due to machine malfunction and failure:

- direct labor idleness, inefficiency, and overtime
- lost production due to machine downtime
- premium shipping to avoid late delivery
- relocation and setup of machine tooling
- idle banked parts to cover machine downtime
- rework and repair of production parts
- creation of production scrap
- product quality problems
- increased cost of warranty

The group was then shown published data from one of the big-three US automobile companies, where it was estimated that for every dollar spent on maintenance labor and materials, there was another three to five dollars in maintenance-related costs. This data is referenced in the business case. Obviously, this process did not have access to the company's closely held financial data. The financial data presented was based on estimates and manipulation to preserve the identity of the company involved.

Using the lost-time data, the following business impact statements were included as a part of the overall findings presented to the plant manager and his staff. It included details about how the workshop was conducted and how the estimates were translated into maintenance costs, maintenance-related costs, and the total cost of maintenance. This information was presented as follows:

- **Business Impact Statement #1**
  (assuming 250 skilled trade employees @ $50,000/year)
  Maintenance labor = $12.5 million/year

- **Business Impact Statement #2**
  (estimated 35 percent of maintenance labor is wasted each year)
  Wasted labor is equal to $4.375 million/year

- **Business Impact Statement #3**
  (maintenance expense directly impacts profit)
  Profit reduced by $4.375 million/year

- **Business Impact Statement #4**
  (estimated 35 percent of maintenance labor is wasted each year)
  Wasted labor is equal to 88 full-time employees

- **Business Impact Statement #5**
  (assuming average cost to build a semi-tractor = $55,000)
  Maintenance labor losses equal to 80 semi-tractors

- **Business Impact Statement #6**
  (assumes a 5 percent profit)
  Maintenance labor losses equal to $4.6 million in sales

- **Business Impact Statement #7**
  (assumes labor/material ratio to be 1:1)
  Maintenance labor and material = $25 million/year

- **Business Impact Statement #8**
  (based on a published estimate by one of the big-three automobile companies that they spend three to five dollars on maintenance-related costs for every dollar spent on maintenance labor and material; in this case, $3 in additional maintenance-related costs is used)
  The total cost of maintenance = $100 million/year

If the $100 million estimate above triggers an emotional knee jerk or an out-of-hand rejection, relax! The accuracy of these estimates or the math involved isn't the real issue. Although these plant-maintenance cost estimates may seem mind-boggling (and even if the estimates were off by 50 percent or more, and they probably are not, the numbers are still mind-boggling), this is not a reason for alarm or panic. This is a wake-up call! This situation didn't happen overnight. It has evolved over time, slowly growing worse over the past fifty or sixty years, keeping pace with the ever-increasing costs and complexity of discrete manufacturing.

What's most important about this exercise is whether or not it successfully conveys to top management the enormity of the total cost of maintenance and the ultimate bottom-line impact it has on the success (or failure) of the larger business enterprise. If the problem seems a bit overwhelming, there's a saying that "problems are simply opportunities turned inside out." In that case, the magnitude of the problem presents an opportunity of similar magnitude.

Based on the author's personal observations and experience during forty-plus years in plant maintenance operations, the reason for these problems has become obvious. As Edward Deming would have said, "This is a system problem, not a people problem." It's not a matter of who is responsible, it's a matter of *what* is responsible. The *what* is all too common in US manufacturing plants: dysfunctional management controls. Dysfunctional management controls become the underlying root cause of excessive maintenance costs and losses.

These controls are put into place by management. They are management's tools, and fixing the problem is management's responsibility. In fact, it cannot be done without their involvement.

Now for the good news. In situations similar to the business case presented here, a reduction of 20 to 30 percent or more in maintenance-related costs and losses, plus a 10 to 15 percent reduction in maintenance department costs, is possible over time using the concepts of synchronous planned maintenance. Significant savings are possible within the first year; maximum savings are possible within a two-to-three-year time-frame. Remember, these savings continue year after year.

Obviously, the ultimate savings, and the time required for reducing these costs and losses, are dependent upon adequate planning, proper execution, and how aggressively the plant pursues the processes of constraint management. When properly designed and implemented, the concepts of synchronous planned maintenance will help to minimize the total cost of maintenance for the plant. It's all a matter of properly engineered constraint management, and that's what SPM is all about.

Synchronous Planned Maintenance

MISSION

To ensure quality in every aspect of preserving the plant's physical assets.

VISION

Plant maintenance will become an enabler of the goals and objectives of the customers it serves.

PHILOSOPHY

Utilize the knowledge, skills, and talents of everyone within the business enterprise who can help in bringing about positive change.

OBJECTIVE

Minimize the total costs of maintenance, thereby reducing the cost of production.

METHODOLOGY

Remove or reduce the constraints, roadblocks, barriers, and delays that maintenance skilled-trade workers encounter every day in their efforts to minimize the adverse effects of machine malfunction and failure.

GOAL

Given full management support and resources, a 25 to 35 percent reduction in the total cost of maintenance is achievable within twelve to eighteen months.

About the Author

Robert S. Hilligoss began his forty-two-year career at General Motors as a skilled tradesman in the maintenance department. He also served as a maintenance foreman, general foreman, and as a maintenance specialist in plant engineering. Later, in his career as a manufacturing consultant at Electronic Data Systems, he provided both maintenance training and consulting support to GM plants.

www.ingramcontent.com/pod-product-compliance
Lightning Source LLC
Chambersburg PA
CBHW081429220526
45466CB00008B/2323